Twilight of the Anthropocene Idols

Critical Climate Change

Series Editors: Tom Cohen and Claire Colebrook

The era of climate change involves the mutation of systems beyond 20th century anthropomorphic models and has stood, until recently, outside representation or address. Understood in a broad and critical sense, climate change concerns material agencies that impact on biomass and energy, erased borders and microbial invention, geological and nanographic time, and extinction events. The possibility of extinction has always been a latent figure in textual production and archives; but the current sense of depletion, decay, mutation and exhaustion calls for new modes of address, new styles of publishing and authoring, and new formats and speeds of distribution. As the pressures and re-alignments of this re-arrangement occur, so must the critical languages and conceptual templates, political premises and definitions of 'life.' There is a particular need to publish in timely fashion experimental monographs that redefine the boundaries of disciplinary fields, rhetorical invasions, the interface of conceptual and scientific languages, and geomorphic and geopolitical interventions. Critical Climate Change is oriented, in this general manner, toward the epistemo-political mutations that correspond to the temporalities of terrestrial mutation.

Twilight of the Anthropocene Idols

Tom Cohen, Claire Colebrook, J. Hillis Miller

()
OPEN HUMANITIES PRESS
London 2016

First edition published by OPEN HUMANITIES PRESS 2016

PRINT ISBN 978-1-78542-015-3
PDF ISBN 978-1-78542-016-0

OPEN HUMANITIES PRESS

OPEN HUMANITIES PRESS is an international, scholar-led open access publishing collective whose mission is to make leading works of contemporary critical thought freely available worldwide. More at http://openhumanitiespress.org

Contents

Preface

Tom Cohen and Claire Colebrook

When we conceived this volume the original title was *Twilight of the Anthropocene.* We liked the Nietzschean resonance, and the notion that by the time the Anthropocene arrived as a marked event, yet still to be consecrated by geologists, it was already far too late. By this we do not mean "far too late to save the planet." The *planet* does not need to be saved; it existed before organic life, and will go on to exist for some time (probably) well after humans and well after organisms. The "lateness"—or *twilight*—is not even a lateness for us. Indeed, one of the features of what has come to be known as the Anthropocene is that very few want to own up to being the guilty party: as soon as the Anthropocene was declared as a way of uniting humans once again, objections started pouring in. Why would "we" want to sully the entirety of humanity by placing it as the author or agent of this late-modern event? So many declare, against the Anthropocene: "Not in my name!": the Anthropocene is really the Capitalocene, the Corporatocene, *or* is better figured as a critical zone rather than one grand evil mess that includes all of humanity. And so the first way we would like to intertwine the concept of the Anthropocene with the notion of twilight is to argue that the idea was at once highly illuminating, suddenly sweeping away all concepts of the post-human, erasing the fiction of Cartesian "man," and allowing humanity to appear clearly and distinctly, and yet the blinding nature of this light obscured so much. Only now, as the dazzling brilliance of the Anthropocene idea begins to wane, and we hear all the claims for different scales and narratives, do we perhaps think to question *both* the logic of "anthropos" as the single agent of geological change *and* the cry from the other humans who accept the narrative of geological destruction but want to exempt themselves. It's capitalism and corporations, not *me,* not *my humanity.* If the Anthropocene seemed to drown out other scales and figures with its blinding light, its dimming seems to have opened other narratives, and yet perhaps what is not questioned is the light of narrative as such.

Everything appears both with conditions of visibility, but also of obscurity, and one might only become aware of constitutive blindness by way of another dimming of lights. (The "light" of reason progressed only by co-opting fire, coal, nuclear energy and the labor of many beings not blessed with the spoils of enlightenment; but reason can be made aware of that debt only when the source of light is waning, in an age of depletion.) Insofar as there is a "we" or an "us," we cannot say, in good conscience, that we only found out that we were destructive once it was too late. The formation of a "we" is generated from destruction and from the recognition of destruction: humanity as global anthropos comes into being *with* the Anthropocene, with the declaration that there is a unity to the species, and that this unity lies in its power to mark the planet. To speak of twilight of the Anthropocene has a three-fold resonance: the concept is already waning precisely because the bet it placed on naming "Anthropos" once and for all has met with so many objections (usually objections from those who want to save humanity from the charge of global destruction) that seeming counter-narratives are being written (with capitalism, colonialism, patriarchy and corporate culture being offered as more nuanced ways of naming the component of the earth that has become the agent of systemic change). Second, this play of lights within the global narrative—of who, and when, and how—obscures the light of narrative as such: both the Anthropocene and its competitors assume that the globe as a living system can be marked at certain points, and that these points *are to scale.* To question scale as such by way of the concept of twilight is *not* to reduce these narratives to human construction but rather to place the human and narrative within *syntheses* (of before and after, the earth conceived as a globe and relatively enclosed space), and to acknowledge that the forces from which various scales and narratives are proposed are multiple and irreducible to any register. To illuminate is to (at least in part) occlude. Finally, in a far more parochial manner, if we write of Anthropocene idols and twilight, this is not because we set ourselves apart as critical dragon-slayers, tearing down the great idols of theory and humanism to arrive at a properly post-human apocalypse. On the contrary, we find ourselves still conversing with the critical names and manoeuvres of the past, and yet for all their power and for all the illuminating force of their declarations, belief in the world is at least as strong as it ever was. If, decades ago, Jacques Lacan posited the real as that which resists symbolization absolutely (Lacan 1991), and Paul de Man (1996) argued that there could be no theory of narrative (for such a theory would be a narrative), and Luce Irigaray (1985) argued that figuration of the scene of truth was always speculative (generating some material substrate that would be the basis for the subject and knowledge), these attempts to hinder the Olympian self-regard of what defined itself as man ultimately resulted in reaction formations. Not only has the narrative of humans as a destructive species generated the imperative to survive—if "we" discover ourselves to be an agent of destruction, then "we" must re-form, re-group and live on; the very critical motifs or idols that offered another way of thinking about the future became the means for a hyper-humanism. Somehow Lacan has enabled Slavoj Žižek to hold onto the motif of genuine revolution; a notion of deconstruction as textualism has allowed for various turns back to affect, matter, bodies and realism, and Luce Irigaray has found a new future for humanity by way of the East: "To go back and meditate starting from practices and texts of Eastern cultures, especially pre-Aryan aboriginal ones, can show us a way to carry on our History" (Irigaray 2002, 36).

One of the unifying motifs across this volume, that all three of us explore in different registers, is that there is no "we," no "anthropos" until, in a final moment of inscribed and marked destruction, a species event appears by way of a specific geological framing. When we began writing this volume some years ago, the Anthropocene was a relatively fresh notion, and seemed to promise—even if it was the *Anthropo*cene—some sense of a new modality of theory. This could either occur by way of generating viewpoints, framings or intuitions of an inhuman look, or of refusing the inscription of "the" human altogether. When, decades ago, Thomas Nagel posed the question "What is it Like to be a Bat?" he concluded that no matter how much information we might gather about bat behavior, experience, navigation and sense input, we could never *live* the bat in his batty world (Nagel 1974). Our slightly different project was to ask, "What is it Like to Be a Human?" and 'What is it Like to think without the Human?" Our various answers were no more generative of lived experience than Nagel's. We can summon all the information we like; we can shift scales, read more and write more, but this reading writing animal, finding itself inscribed in the Anthropocene, cannot exit from inscription altogether and simply live:

> Observe the herd as it grazes past you: it cannot distinguish yesterday from today, leaps about, eats, sleeps, digests, leaps some more, and carries on like this from morning to night and from day to day, tethered by the short leash of its pleasures and displeasures to the stake of the moment, and thus it is neither melancholy nor bored. It is hard on the human being to observe this, because he boasts about the superiority of his humanity over animals and yet looks enviously upon their happiness—for the one and only thing that he desires is to live like an animal, neither bored nor in pain, and yet he desires this in vain, because he does not desire it in the same way as does the animal. The human being might ask the animal: "Why do you just look at me like that instead of telling me about your happiness?" The animal wanted to answer, "Because I always immediately forget what I wanted to say"—but it had already forgotten this answer and hence said nothing, so that the human being was left to wonder (Nietzsche 1995, 87).

When Nietzsche wrote about humans looking longingly at animal forgetfulness he was more prescient even than his declarations of being untimely would promise. With all the evidence of human destruction—or the human as that which finds and inscribes itself after destruction—it seems as though only an animal can save us, as though it might be possible now, finally, to become-animal. When Brian Massumi looks to the future by detailing what "we" can learn from animals about politics, he suggests a program of *replacing* the human, back to its animal milieu:

> This project requires replacing the human on the animal continuum. This must be done in a way that does not erase what is different about the human, but respects that difference while bringing it to new expression on the continuum: immanent

> to animality. Expressing the singular belonging of the human to the animal continuum has political implications, as do all questions of belonging. The ultimate stakes of this project are political: to investigate what lessons might be learned by playing animality in this way about our usual, all-too-human ways of working the political (Massumi 2014, 14).

Our volume takes a different path from that of talking to the animals, or of repairing what is left of the human. Tom Cohen's genealogy and geology of the *anthropos,* shifts terrain from the problem of whether we—we humans—can legitimately know or feel what it is like to be other than human. There is no shortage of talk and lament regarding the human, and both its incapacity and capacity to think outside its all too human subjectivity. But all that talk about the post-human, the non-human, the inhuman and the problem of lumping all humans into the Anthropocene provides a way of sustaining the human as a problem. What if the human were an effect of its own delusions of self-erasure? What if there were no humanity other than that which is effected from the thought of the other-than-human? We can think of this in many ways. One way would be to see the constant proclamations of overcoming humanism, Cartesianism and anthropocentrism as *producing* man as the being who can annihilate himself in order to become animal.

Tom Cohen's approach in this volume is to reverse the problem of anthropomorphism: it is not so much that "we" project "our" humanity onto nature, but that there is no "we," no humanity until we chastise ourselves for configuring nature in our own image (which does not exist). The mourning of humanity, the accusations that "we" have not attended sufficiently to our inhuman others, the extension of human rights or personhood to nonhumans: all these Anthropocene gestures are modes of generating a humanity that never was. Following Lacan's declaration that there is no sexual relation, Joan Copjec asked us to "imagine there's no woman" (Copjec 2002). Timothy Morton has also declared that there is no nature (Morton 2007). The lost, prohibited, yearned-for object of fulfillment does not exist outside structures of mourning. Today, we do not want to say "imagine there is no humanity"—no all-encompassing unified species unity charged with the crime of the Anthropocene—for that is the gesture of post-humanism, of thinking beyond the human. Rather, there never is and never was *the human,* until perhaps some pious theorists thought there might be a beyond of the human or a humanity to come. Humanity comes into being, late in the day, when it declares itself to no longer exist, and when it looks wistfully, in an all too human way, at a world without humans. The human is an effect of a declaration of non-being: "I do not exist; therefore I am."

It is in this respect that Hillis Miller will take up the notion that climate change is a linguistic event: just as one might say that *the* human is an effect of strategies of thinking the other-than-human, to say that climate change is ideological or linguistic is not to say that it isn't real. It *is* to challenge the way one thinks about the morality, temporality and rhetoric of reality. As Bruno Latour argued, one of the great successes of the climate change deniers was the promulgation of a notion of a true reality that we might grasp behind all our figurations, twinned with the notion that science should ideally be ideology-free. Climate science, like all science, is composed and compiled from a series of dispersed institutions and practices (Latour 2014). But there is no *reality* that we should try to grasp if only we could somehow get outside inscription or composition. Further, and this is where Hillis Miller draws on de Man's conception of ideology: rather than see ideology as the constructed, linguistic, symbolic or cultural mask that conceals political reality, it is the notion of some pristine real prior to inscription that is ideological. At first this might seem to play into the hands of climate change deniers; for if this is so, then how would we ever *know* once and for all that climate change was anthropogenic? Your guess would appear to be as good or bad as mine. But *that* notion—the notion that science is as much a construction or rhetorical device as anything else—is a weak, and still ideological notion. If inscription goes all the way down, and if there is no nature, no climate, no humanity and no truth that would exist outside inscription, then what remains is the *reading* of inscriptions. This remainder operates at two levels. First, as Hillis Miller notes, "Language is deeply involved in this happening. It has been facilitated in part by climate change deniers, who believe the lies told by politicians and the media claiming that the evidence for 'anthropogenic' climate change is a hoax perpetrated by 'mad scientists,' against the evidence." To talk about language, inscription, rhetoric and tropes is *not* to introduce textuality into an otherwise scientific or material problem; every fact, every reported event, every filmed disaster, every declaration of drought, every discussion of just where the Anthropocene starts and stops is already within language. And language is material in a radical sense: not the medium through which thought communicates, but a multiplicity of relations and traces that enables what comes to experience itself as thought. In addition to the irreducible materiality of inscription and its resistance to being thought, there is a second more profound problem of inscription, which all three of us work with in writing this volume.

To talk of inscription is not just to talk about language or even visual composition in its narrow sense; it is not only to concur with Latour that any scientific account is composed from technical readings, adjustments, concerns, interests and affect. Rather, it is to see the world, the earth, the climate—all these unities that *we* are witnessing as being changed utterly—as effects of complex systems of relations that are irreducibly multiple. As Stephen Jay Gould noted decades ago, "we" have a parochial interest in the survival of our species (Gould 1998). But one might go further, and say that *there is no we,* only a network of parochialisms, composing now this and now that sense of urgency. To talk of saving humanity, or saving the planet, or even being past *the* tipping point is –necessarily—to be perceiving a certain time and space from a certain scale, vantage point and network of syntheses. The tipping point was probably reached centuries or millennia ago for some living beings whose carnage was required in order to generate the "human" civilization that is now mourning its own demise.

Our original intention, then, was to think about the Anthropocene as a twilight concept, as a form of half-recognition that can only occur in the moment of waning. And we also wanted to signal that what appears as a moment of sudden loss or intrusion—"look, we destroyed the planet! Who would have thought!"—was there all along. There was always destruction, always eco-cide, but "now" (for some) it has become readable (even if, for others, such destructive force was all too obvious, and human, all too human).

Colebrook's angle of entry is entirely through the orders of the aesthetic—which is to say, the materialities of inscription that, inaccessible to perception or memory regimes, give rise to perceptual grids, "phenomenalities," the cinematic hallucinations and aesthetic ideologies that have, in the case of *anthropos*, shaped not only the modes of reference, hermeneutics, and visibility that have accelerated the ecocidal logics of any anthropocene beyond "tipping points," but the problem of reading that subtended this arc. The "aesthetic" is called back to primacy of agency around the inverse way that the figure of the sublime had inhabited, recuperated, diverted, or implicitly doomed the Euro-anthropic traditions of Western writings from which it derived.

In each of these explorations, the "idols" that we cast about involve what has been ignored or largely omitted from the chorus of facts, corporate media white-outs, green or sustainability imaginaries, and diversionary activisms attending the topos of "climate change." What has been suppressed is that language conventions and cognitive habits have abetted or predetermined the accelerated mass extinction processes under way—irreversibly so. If the "man" of anthropos was a Greek invention, he appears with the very definition of the polis or the political that, for mysterious reasons, "we" are still trying to recover or re-inflate. The paralysis of the liberal and utopian left before the logics of climate change mirrors that of the humanists. This volume, accordingly, extends the cipher for reading the *neganthropocene*, today, back to where these terms seem, alone, to have been addressed in the twentieth-century "theory" canon: the work of Paul de Man. Our previous collaboration, *Theory and the Disappearing Future: On de Man, On Benjamin* (2012), explored this problem by extracting a "de Man" who appeared, from a twenty-first century perspective, to be writing of, or to, the "Anthropocene" *avant la lettre*—something quite distinct from the narrative of "deconstruction" and its self-mythologizing and branding. In resetting that narrative, it would appear that Derrida operated almost as a climate change denier, and in trying to cement his own legacy, occluded a materiality that would put the premise of "deconstruction" in doubt from within. By contrast, de Man went toward this locus.

In this current volume "de Man" is honed into a cipher for the following question: if the passage today into an ecocidal script delegitimizes the forms of archival management, governance, and epistemologies that operate today, what do the materialities of inscription, the "epistemological critique of tropes" or "aesthetic ideology" more broadly have to do with this ecocide? In many different ways, this volume approaches the passing of tipping points as a positive achievement for thought, and the irreversibility of ecocide as the invitation to a terminological reset of "idols."

All of our chapters, in different ways, depart from where our previous volume terminated: each begins where our first explorations left off by asking what tools might be available for a post-mortem of the Anthropocene as a rhetorical term or claim.

~

As the writing of this book progressed, it was not so much the Anthropocene that was catching our attention as the Anthropocene idols: not just the disaster-mongers, emergency opportunists and ecocide impresarios who could market survival strategies, but the theorists who thought to find a new point of refuge. In its most explicit form one might think here of the criticism made by Naomi Klein of the forms of capitalism that profit from catastrophe, but one would also need to think about Klein's own redemptive narrative in the same mode (Klein 2014). Only now, in this moment of "final" catastrophe, can humanity save itself from a capitalism that is now—finally—appearing at its most destructive and unsustainable. Turning to the *Avatar* solution (as typified by James Cameron's 2009 eco-redemptive epic that retrieves holistic indigenous wisdom) and the imaginary of a return to 60's social movements, the crashing of planetary life has one great virtue: it confirms the critique of twentieth-century socialism (even as the latter fed the same acceleration). It is interesting that the author of *The Shock Doctrine*, expects a "people's shock" to stir the cognitive cavities of the multitude in a non-violent and inclusive "revolution" that would have to be so thoroughly cultural, systemic, cognitive, and sudden as to exist only as one of those "people to come" that political discourse, if it is that, rallies itself while deferring.

The complicities that Klein's maneuver displays operate far more widely in a series of objections to the Anthropocene itself. It is as though only with the heavy-handed humanity-accusing declaration of the Anthropocene, could theory and the humanities revive itself by talking of some guilty humans who deprived us of our just and savable world. One might suggest that an older gesture of deconstruction—that all our concepts of justice and democracy *in their very corruption* must promise a humanity to come—has now gained much broader purchase, even as a covert divestiture of responsibility. If "our" planet, and "our" humanity is broken then there is necessarily the promise of another humanity and another future, still. We see this at the end of all Hollywood's Cli-fi fare. Here, though, is where one might channel Nietzsche, for whom the *twilight of the idols* is constitutive of a certain philosophical piety. It is not that master thinkers (or humanity) have fallen on hard times, but rather that the very declaration of hard times—of crisis, emergency, decadence, loss of reason, injustice–enables the pious elevation of the master thinker.

For Nietzsche, a certain demand that the future is promised, and promised *to us,* follows from an experience of decadence. Reason, history and a promissory arrow of time are both modes of recuperation, of refusing to experience the world as not necessarily *for us,* as not *necessarily* on a path of human moral improvement at all. Against this necessary sense of promise, a sense of history would entail a consideration of inscription, and of the contingent composition of order that allows narratives of causality, of relation, and of justification to range.[1] The history of histories is just this war of inscriptions, but with this caveat: this history, and its sense of being *for us,* is always cobbled together after the event:

> Language began at a time when psychology was in its most rudimentary form: we enter into a crudely fetishistic mindset when we call into consciousness the basic presuppositions of the metaphysics of language—in the vernacular: the presuppositions of *reason.* It sees doers and deeds all over: it believes that will has causal efficacy: it believes in the 'I', in the I as being, in the I as substance, and it *projects* this belief in the I-substance onto all things—this is how it *creates* the concept of 'thing' in the first place ... Being is imagined into everything *–pushed under everything*—as a cause; the concept of 'being' is only derived from the concept of 'I' ... In the beginning there was the great disaster of an error, the belief that the will is a thing with *causal efficacy,*—that will is a *faculty* ... These days we know that it is just a word (Nietzsche 2005, 169).

Following Nietzsche, it is in the contemporary experience of reason's failure—of the world being at odds with reason, hope and piety (or humanity in general)—that reason becomes evermore shrill: there must be a future and it must be human (and by human, we mean "mine," ours, and not some other unimaginable life or non-life). The "twilight of the anthropocene idols" does not refer to a loss of reason, of critical thinking, of our human potentiality. On the contrary, the very figure of a humanity oriented towards a history of flourishing, self-realization, universal scope and a proper future relies upon an accidental and temporary corruption. For Nietzsche, a dream of human futurity (always better, always more human, more rational) is an eternal idol that relies upon declaring its own threatened existence: "I am becoming extinct, therefore I am, *and ought to be.*"

> For thousands of years, philosophers have been using only mummified concepts; nothing real makes it through their hands alive. They kill and stuff the things they worship, these lords of concept idolatry—they become mortal dangers to everything they worship. They see death, change, and age, as well as procreation and growth, as objections,—refutations even. What is, does not *become*; what becomes, *is* not... So they all believe, desperately even, in being. But since they cannot get hold of it, they look for reasons why it is kept from them. 'There must be some deception here, some illusory level of appearances preventing us from perceiving things that have being: where is the deceiver?'—'We've got it!' they shout in ecstasy, 'it is in sensibility! These senses *that are so immoral anyway,* now they are deceiving us about the *true* world.' Moral: get rid of sense-deception, becoming, history, lies,—history is nothing but a belief in the senses, a belief in lies (Nietzsche 2005, 167).

More specifically, as outlined by Hillis Miller, one can observe that an entire rhetoric of lies, hoaxes, manufacture, ideology and concealment has allowed climate change denial to survive. Whatever is experienced—whether it be scientific data, an absence of rain, freak storms—it is possible to posit a deeper truth, behind these lies: that climate change is the grand conspiracy fiction of the left. If we insist here that climate change, financial brigandry and eco-cide are real *and* inscriptive events this is because we are committed to reading materialities. [2]

The truth is composed from these inscriptions—with the inscription of sea levels, extinction rates, conspiracy theories, dreams of revolution, post-apocalyptic reveries, dire warnings, and geo-engineering proposals all being material events.[3] What we insist upon is that rather than talking about what these inscriptions must mean for us, we should read: do not see the snow storm as a sign that there's no global warming, as it does not erase or allow us to read as secondary all the other matters at hand (droughts, floods, resource depletion, extinction); we do not see a rise in employment as a sign that there was a brief recession that must (like previous recessions) return to normal. Do not read catastrophe as a sign that there must be a humanity to come. More importantly still, perhaps if one began to read catastrophe—rather than fold it about one's own person, world and temporality—one might have to confront a radical temporality, in which what comes to pass might not be in the order of history. In this respect one might draw upon Michel Serres's conception of history as a strata of inscription, where certain lines, marks, events and orders initiate relations among traces that will proceed *until* one reads and imagines *not* a time of progress but a sublime becoming (one not amenable to the line of time as we know it). In his book on Rome Serres depicts the history of a space as a confluence of contingent inscriptions—akin to the emergent creation of a termite hill. In the beginning is a random collision that acts as an attractor; far from seeing this line of catastrophe as promising a justice to come, the task of reading is one of retracing towards contingency, with each step back giving nothing more than marks and reversals:

> Imagine the ground of Rome after a millennium of trampling by the Romans. Imagine the earth of the forum after the pounding of the feet of the mob. And now, decipher that ichnography. This is the final painting of the Herculean meadow, this is the initial painting of Rome; these paintings have prescribed every direction or meaning. There is prescription of every direction or meaning before the inscription of a single direction or meaning. In the beginning is the ichnography.

That is to say, the integral, that is to say, the sum, the summary, the totality, the stock, the well, the set of meanings or directions. The possible, capacity. Each defined direction or meaning is only a scenography, that is to say, a profile seen from a certain site (Serres 2015, 30).

Chapter One

Trolling "Anthropos"—Or, Requiem for a Failed *Prosopopeia*

Tom Cohen

> *Another form of convalescence, which I sometimes prefer, is sounding out idols. The world has more idols than realities: this is my "evil eye" on the world, this is my "evil ear" as well. Posing questions with a hammer, and perhaps, hearing in reply that famous hollow sound that indicates bloated intestines—what a pleasure for someone with ears even behind his ears.... Even this work—the title gives it away—is above all a recuperation, a sunspot, a little adventure into a psychologist's idle hours. And perhaps a new war, too?*
>
> Nietzsche – Preface, *Twilight of the Idols*

1. Daybreak in the Ideovomitorium

Again.... ? Is there time for another replay, or too much time, in the non-now of 2015—that is, as the geo-political and geo-rhetorical apparatuses compensate for what may take years, still, to enter the normative imaginary: that those tipping points, the ones that the temporal imaginaries of ecological protest want to outrun, pre-empt, save from, or mutate with, have passed, tripped, entered the phase of vortex-like accelerations and back-loops, taken any "decision" away from us, introduced something like a politics of managed extinction going forward (already visible)? It would be the year, or date hypothetically, to which the geo-morphic and bio-morphic scar of a human or Anthropocene age would be, by some future archivist, indexed. Since the title of this volume was to pivot on

"Twilight of the Anthropocene" or "the Anthropocene Idols," one enters an operatic *mise en scene*. Even when Nietzsche uses this imprint, it converts (and of course tweaks or mocks) Wagner, draws on that tired use of "twilight" to romanticize a passing, or passed, epochality, as it always seemed. Here, in what we will call the Ideovomitorium, recovering from a sort of *Walpurgis Nacht* in which an irreversible calculus has slipped or passed—let's just say, or call them broadly, ecocidal *tipping points*—there is no twilight really to mourn. Waking into the new day, nonetheless, seems disorienting, sunless, encrusted from a long night of cognitive dissipation.... Why, rather than drone on about "twilights" of the Anthropocene, or more strictly, its cognitive "idols," would one not want to rush, first, to the daybreak beyond that?

Nietzsche's evil ear with "ears even behind his ears" in the above quotation would entertain itself by sounding out tinny and hollow idols—a "recreation" and by-play from the task of transvaluating all values. This may all sound somewhat different if we update ourselves from the noon of Zarathustra—where a hyperbolic logic is turned against the "human" as previously constructed, before what is called, sometimes, a responsibility to an "Earth." That hyperbolism may be inverted in the hypobolisms of today. That is, of that other noon moment, today's twilight of the "Anthropocene" where we witness the passing of tipping points, in which a broader ecocidal extinction event is accelerated, and this before a "last man" global tele-culture in which this monumental moment is registered with a blink (if at all). One might think that this passing would draw hysterical philosophic or media attention. The proactive project of the early responses to "climate change"—that of the ecological thought, the holistic and organicist reflexes—has given rise (if not fully passed to) what we would have to call the ecocidal sublime. What are the idols of "the anthropocene," then, today, if not the proliferating cognitive theatrics of what one must call, finally and with some relief, *climate comedy*? There is something comic about today's other or new noon—that, say, of "peak human" (or Human 3.0), passing unnoticed in the tele-epoch of the last man. —*Ah Nietzsche, naif. So far as twilights go, you had it easy—the passing of cognitive empires, the noon or shadowless moment of Western self-organization or regard, the locating of a target.*

The shadowless *noon* of Zarathustra is parodically inverted in today's impassivity of last man settings when "climate change" arrives, and the ecocidal races are on. They are on now. They are also, in a sense, over before arrival. So a very small hammer comes out again, this tool of tools, when much seems exposed before the irreducibility of extinction events, which delegitimize every political organization and epistemography to date. Before this trope of "climate change," discursive practices appear checked, exposed as rhetorical regimes intertwined with cognitive spells, feedback loops, short-circuits, end-running, a gaming of cited "laws." One might grow to suspect that climate change had something, after all, to do with epistemographic regimes, unstable conventions of reference? Clive Hamilton reminds us that an entire trajectory of utopian thought on which "we" rely, was not only a parenthesis of post-war Western hyper-industrial affluence, now withdrawn, but itself fueled the same ecocidal acceleration [2012]. One must develop a taste, here, for *aporia*—and tune out the flabby chorus of last man reactivism (comedic) for whom the thought, today, of extinction events is deferred as something *cynical,* outside the tribe, another excuse. But the reflex of mourning was not for "us," despite its ejaculation and fetishization: one has always wanted to restore what was never there entirely, *anthropos,* who seems to return today in ungainly contortions and mimic memory. Perhaps mourning was for the *biotica* wiped out going millions of years forward in what seemed a decade of procrastination and discursive con games, after all the remarkable accidents that brought about this opening: lucky meteor strikes (the dinosaurs), just right positioning near a star ("*the* sun"), diverse lucky catastrophes and inscriptive events that made *anthropos* appear, among other hominid types and cognitive orders.[4]

Language and "climate change": the first non-term dissolves, today, to absorb hyper-technologies of script and data streaming, tele-marketing, mnemo-technics in exponential digital mutations. This relationship between our language conventions, referential practices, and ecocide is set aside generally today for a number of reasons. One gives all the attention to scientific reports and their mediacratic dilutions or political manipulation. One perhaps ticks off the burgeoning if not cascading and interlocked climate trauma sprouting: megadroughts, accelerating melt-offs, shock flooding, heat wave die-offs, then of course climate wars. It should have been obvious, this lack of attention to cognitive language, and represents a certain blind, an unscratchable itch. Even in the most pragmatic sense, key terms and tropes of "the climate debate" have been disastrous, a retinue of flaccid romanticisms or organic metaphors of recovery, sustainability, environmentality, and so on, coupled with soporific and numbingly descriptive scientisms ("global warming"). It was a discourse engineered for failure and exploited as such by corporate media (certainly in the Anglo or Murdoch-news nations). One gags on the ironies: the term "Anthropocene" can only arrive in (or after) the *twilight* of what it names, so it can only anticipate or legitimize itself from a future recognition of it, after a disappearance it implies is accomplished. It projects a proleptic anterior "inscription." It would have inscribed a proper name into crashed life-systems, "earth," mutated materialities, defaced surfaces, exploited genetic codes—an inscription that would have to be read (that is, given recognition) by another eye entirely. It is driven by an aesthetic compulsion. Today's geographers double back to contain the meme they unleashed, as when they seek now to mark "its" start (the traces of nuclear fission, say, after Hiroshima?). It operates as a destroyer of anthropisms. It is a proleptic memorial of ecocide from a back-glance that is not anthropoid at all.[5]

But the term is a bluff that mutates: it is readily appropriable by corporate rhetorics telegraphing a new ethics of adaptation and geo-engineering projects; it erects a Eurocentric trope into a universal—the Aristotelian *anthropos,* male and Greek, possessor of logic and reason, the artifice of the Greek dawn, and so on. It is not a question of there being rather a better name, a Sinocene, or a Gynocene, or a Bacteriacene.[6] Or, if the future looks back from the perspective of artificial intelligence, robotics and digital "life," then "Anthropocene" as a term would be in bad grace—a destructive time of piggish and ruinous organic life, best suppressed. Does not the U.S. "Hobby Lobby" ruling, after all, give not only personhood but personality to the "corporation" entity, this corporeally non-existent supra-organism, accepted by the Supreme Court of the U.S. as having beliefs and feelings that cannot be overlooked? This supra-organism without a body made a clever move overleaping and consolidating "Citizens United" mere endorsement of its "personhood." There are nations that give this recognition to porpoises or dogs, after all. So it expands again from being a "person" to supplanting the entire facade of subjectivity: this time, the corporation is "more human than human," and will fully take over (and evacuate) its host organism. It is religious, against abortion, for human "life"; it is for family values after all. The outline of personhood has now been invaded and displaced, downgrading the organic or old variety—the messy era of ruinous hyper-industrial civilization, anthropoid, best forgotten.

Of course, this brings another inversion and it explains why *we*, rather than being outraged, seem to shrug at the same time. Perhaps the ascription of personhood to corporations might be true, after all. It is not that the corporations are like "us," but that the structure of the "we" sees itself vaguely in the non-existent corporation's disembodied algorithms. The "we," after all, the premise of tribal or caste "personhood," is an artifice independent from any speaker, one that technically "does not exist" (Stiegler 2014, 12).[7] The corporation is mocking human anthropomorphism. The "Anthropocene," if it exists, is not that of a hominization in which "man" mastered technologies to achieve his apex of planetary power. Rather, it is the hominid as undefined full spectrum organism (fingers, hands, cognition, mnemonically driven) through which host *technics* would incarnate and evolve—to the point at which *it* might re-engineer or shed the host organism, thus enabling the species-split recently engineered (in "2015," say), which implicitly accompanies the "breakaway civilization" of a neo-feudal klepto-mediacratic hyper elite. Even the meme of "inequality" is circulated and sported as a distraction.

What are the idols of the *Anthropocene*? Writing from what we call "2015" one has a certain vantage, even a theatrical responsibility, to note some recent turns in the road of time. Those are not, necessarily, the rupture of the international "legal" order, the flaring of wars and collapse of nominal democracies; those are not the proliferation and acceleration of "climate change" debacles (mega-droughts, air-pocalypses, breaking ice shelves, resource wars); those are not the flare-ups of Ebola or ISIS, and severed heads dangled on videos cut by British-born rappers—and so on, a ripe list. Rather, I limit myself to two events that passed unremarked largely. The first is the passive recognition by Western states that the ecocidal "tipping points" have basically passed. This rewrites the "noon" that

Twilight of the Idols invokes in "How the True World became a Fable: the History of an Error":

How The 'True World' Finally Became A Fable

The History Of An Error

1. The true world attainable for a man who is wise, pious, virtuous,—he lives in it, *he is it.* (Oldest form of the idea, relatively coherent, simple, convincing. Paraphrase of the proposition 'I, Plato, *am* the truth.')
2. The true world, unattainable for now, but promised to the man who is wise, pious, virtuous ('to the sinner who repents'). (Progress of the idea: it gets trickier, more subtle, less comprehensible,—*it becomes female,* it becomes Christian ...)
3. The true world, unattainable, unprovable, unpromisable, but the very thought of it a consolation, an obligation, an imperative. (Basically the old sun but through fog and scepticism; the idea become elusive, pale, Nordic, Konigsbergian.')
4. The true world—unattainable? At any rate, unattained. And as unattained also *unknown.* Consequently not consoling, redeeming, obligating either: how could we have obligations to something unknown? ... (Gray morning. First yawn of reason. Cockcrow of positivism.)
5. The 'true world'—an idea that is of no further use, not even as an obligation,—now an obsolete, superfluous idea, *consequently* a refuted idea: let's get rid of it! (Bright day; breakfast; return of *bon sens* and cheerfulness; Plato blushes in shame; pandemonium of all free spirits.)
6. The true world is gone: which world is left? The illusory one, perhaps? ... But no! *we got rid ofthe illusory world along with the true one!* (Noon; moment of shortest shadow; end of longest error; high point of humanity; INCIPIT ZARA THUSTRA.') (Nietzsche 2005, 171).

This is fine as far as it goes, but what if Zarathustra doesn't fill out the script? Benjamin had already posed this question in dissing the *Overman,* and displacing him, simply, with what he termed the *Unmensch*—a figure he describes as that which happily cannibalizes and destroys "men."[8] Today, this "noon" is bio-material: the time of peak human, peak water, peak species, and the acceleration of irreversible ecocide. So, one might channel Nietzsche's *Twilight* to test some idols of the Anthropocene, to scan about without the requirement, generally, to return in some sort of redemptive reflex to save a twentieth-century cognitive legacy or to somehow save "the planet." "2015" interrupts by exceeding the tragic rhetoric warning of global warming, accelerated ecocide, or extinction. I would suggest, rather, that something like *climate comedy* is in order, now that tipping points are in the back mirror. [9]

I have changed my positions, so to speak, on several things (not that those are particularly relevant and I mention them only in candor). One scrambles to keep up with the accelerations themselves, the sheer *tempophagy*. I no longer find the sobriquet and place-holder "climate change" anything more than tinny; yet a temporary replacement, such as *ecocidal acceleration*, is no less so: it merely draws the perspective from an event to a *process* of which the cognitive addressee ("we") is itself a product and accelerant. If I allow myself a sort of *hypo*bole, to the degree that the import of the "resistance to theory" (de Man [1986]) is parallel today to the manifest resistance to "climate change"—to its re-cognition, to a discourse incorporating it, to responsibility—it would be because both converge on the "materiality of inscription" that puts into question the entire cognitive programs of, well, "anthropos." And here another bit of climate comedy returns. To give this a little piquancy, I'll propose a *recreation*, intended with a very straight face, in the following hypothesis: that rather than having arrived at a geo-bio-technic impasse that requires the total retirement of literary interests from the scientific-philosophical-economic stage, one can posit something like a *literary structure to "climate change,"* one that even guarantees ecocide. When I say "structure" I am trying to obey protocols, be timid—but one might almost substitute *cause*. Exploring this would require, for a moment, detaching oneself from the frontier logics encountered today: that the "norm" we want to return to never was a norm, but rather an abnormal parenthesis. Ecocidal thought compels a new (intellectual) ethics, which would re-read the entire archive, in turn, and rewrites geographisms and agency, and so on.

What fades today is what mobilized the ecological thought or, for that matter, the over-late appropriations from the Left (Naomi Klein [2014]). One abandons the premise used so far: that it was ever a matter of scientifically bringing to light an unwelcome truth in order to get a global response that would over-ride corporate media denialism and addictogenic accelerations. But this was naïve: *they* would not stop in any case, but rather accelerate (say, emissions)—after dividing populations from wealth and control of resources, or preparing a technologically enhanced survivor caste for the next generations.

This anticipates a second signal shift or event of "2015," one which future archivalists will wonder over. The current chatter about "inequality" represents a smokescreen or fop in this regard (Thomas Piketty's meme [Piketty 2014]). That is, this timid word suggests a balance to return to, ideally, called "equality," and insinuates the past equilibrium can be recovered. This typically misreads the situation of today—which *must be read from "climate change,"* which is to say, after tipping points, a sort of ecocidal sublime and the backglance of future imaginaries. These imaginaries or calculations are widely known—one might begin with the 2003 Department of Defense report suppressed by Bush at the time, which dismisses any "war on terror" and envisages mega-droughts, mass climate migrations, population culling and perma-war (starting, well, right about now). If one allows that the delay of aid to Ebola-stricken Africa (much remarked on, including by the U.N. Secretary-General) was not entirely innocent, one might say bifurcations are already well in place. One might note the climate comedy of fortress Europe scurrying to save face (and a few boat drownees from North Africa, as if this were not a trickle compared to what's coming), or observe the outright doom cast on the stateless Rohinga—enslaved, drowned, sold, or pushed back to sea with an occasional water bottle as disposables, a preliminary test case in select population disavowal and *becoming waste*. Of course, as with Oedipus, it is fruitless to look for the cause of the wreckage or to defer some coming catastrophe; we are in the middle of it already, or in fact late, in a twilight of sorts, since from the point of view of any other contemporary life form or living system, "we" *are* the catastrophe. This is why it is the term "extractivism," which is more primordial and, arguably, precedes even "Capital" heard or thought as a totalizing evil agency that we resist but are caught in. One also might take into account today that any Anthropocene would be a subspecies of what Derrida nails as *human-ualism*—which includes the prioritization of *the hand* and the *senses* as configured by historical regimes, of which the haptic-optic interface is a key dominant.[10] This gives a hint of what I mean by climate comedy. It suffuses, perhaps, a spectrum of recent Hollywood Cli-fi fare seemingly designed to mop up the climate change imaginary—since they routinely have survivors and new beginnings tacked on (message: "Chill, *we* win").

One must admire the efficiency of what would be the greatest engineered mass wealth transference since Genghis—under the cover of the "2008 financial crisis" (in which the "shock doctrine" is turned by Capital against itself). It is interesting that Naomi Klein, who codified this analysis, is left with no other turn than to argue, before the immutables of climate change that shred any utopist calculus, for a "people's shock"—resulting, optimally, in a new people's movement like that of the U.S. civil rights movement (that is, her memory tapes of heroic victory—ignoring the current state of re-tribalization under climate change pressure). One cannot do away with the imaginary of the political "we," yet what if the problem, so to speak, lay elsewhere: in the machinal hypotheses of the "we," as practiced (always human on human), tribally posted, exclusionary, inventing the closure that, briefly ameliorative for positing an authority of the imaginary "collective" but structured, already, like and as a corporation—and corporation as "person" specifically. What if there were no outside of this impasse, today, none at all, because any discourse of resistance is both generated or absorbed? This too would lie with how *referents* were contracted, generated, back-looped: the result would be the "prison-house" of short-circuited referentials, memes, hashtags, end-running with algos, derivatives.

To open this portal to take a look occurs in a *lightless* place, which is nonetheless not to be considered joyless or scary—since it precedes Western tropologies of "light," heliotropism (Plato), the Enlightenment, the memory regimes by which something like the eye or seeing were conflated with knowing in Plato's early card trick of fusing the two in the verb *eidein*.[11] And there are two new rules, to surveying these re-organizations of "the inorganic." First, not just that tipping points are past and "we" can now pause to absorb the implications of irreversibility—which alter, in essence, how the entire archive of scripts have been read or transmitted—but that *everyone knows* this. Most aware of all are the hyper-elite and so-dubbed ".001 %," who implicitly fund and tolerate the mass anaesthesia of corporate denialism for the supposed public, while vacuuming up and locking down the future resources that would define survival in a generation and more. *Everything "knows"*—from bio-organisms in mutation due to global warming, to extincting life forms, to migrating species. And yet, we "know" in a way that includes all variations of occlusion, denialism, affirmation, dissimulation, and acceleration. This marks vast portions of our pre-occupations and consciousness. Since social and ideational exchange tends to be a management of conscious hypocrisies, this produces a generalized *climate change "unconscious"* that can be read everywhere, at all points, less stuffed into individual psyches than played out in the still public spaces marked by occlusion. One must discard the Enlightenment tropology, of course, which assumed that "climate change" was a shocking secret that science, bringing to light, would foist on societal governance to address—as if "enlightened self interest" were the market's desideratum. Yet this precludes unrealistically the alternate scenario: not that a collective humanity would come together against a shared threat, as if an asteroid were to strike collectively (though one can imagine first checking where and what cities, reflexively, before signing up, not to mention the e-casino economy's response). Rather, one might find the opposite—a reassertion of competitive groups and tribalisms, virtual combat of non-organic "species" (military-telecommunications systems, corporate networks, political constituencies, "human" types and a breakaway civilization enhanced, today, by different access to resources, financial mafias, proprietized science, bio-technologies). It has become easy to point to something like the hyperbolics of "propaganda" within media systems (the U.S. very much on spot). Each would be without "legitimacy"—since ecocide and accelerating climate change delegitimize received governmental structures. One can crystallize propaganda as a targeted discourse for mnemonic engineering, the insertion of new repetitions (Potemkin inscriptions). Yet one defers concluding a parallel delegitimation of cognitive and hermeneutic settings. Entering a supposed "internet of *things*" brings a double-inflection here. The *telepolis* absorbs and disconnects, totalizes and pretends to seal itself off. It manages the cognitive traffic of tropological nervous systems and memes. But "climate change" is a stand-in term for a radical materiality that eluded the Western epistemological systems and "dialectical materialist" offspring. Everything appears caught in the re-iterations and vibratory shaking of this tectonic impact, exposing discourse as what it had always been—varying forms of power rhetoric shaped and instituted in relation to technologies of memory, knowledge bases, and theft. There is certainly room, given the abyssal performance to date of the reigning trance, to question and dissolve the rhetorical premises of these sentient locusts expropriating the last terrestrial life-systems and resources in quest of energy. Is one of the blindnesses, today, the grab-bag of twentieth-century tropes and critical legacies still regurgitated in order to clothe and manage, name or manipulate, disguise and game their disappearance—where something like "man" has become and performs like the unliving, the "death machine" that Chris Hedges accuses corporations of manifesting, theologically considered?[12] Moreover, anyone wanting to follow, hail, or claim a meme of *justice* has, it seems, to prioritize "social" justice—human on human "ethics," purportedly—and must ignore that a certain justice can be perceived as turning against the entire global Anthropocene "culture" and Last Man passivities and corruption, thus putting in question the sacred composition of the "social," the "we," the utopist together with the cognitive and financial ("fascist") cartels at the other end of an awkward and shared spectrum.

To borrow from Bill Maher, there are now "new rules": regardless of the supposed opinions, spin, strategies of interest, white-outs, capture of affect and attention, and so on. First, *everyone knows*—and, moreover, all so-called ideologies, religious investments, academic consignments, and prognoses register as the literary agents caught in rhetorics of power, reference, stupefaction (overload), and perceptual trance read back, as it were, from the logics of climate change. This might even call for a return to certain surgical insights into the epistemology of tropes and their relation to ecocide—that is, if we view the "Anthropocene" as the latest attempt to name a human trajectory beginning with its initiation, the theft of fire, the *technogenesis* of archival regimes and *we's*. Second, having passed tipping points a "new" or perhaps prehistorial template emerges in which differential questions arise, no longer subjugated to projects of recuperation. Third, "we" have entered into a zone that must be explored, but does not return to the mother ship this time, nor add to the imaginary toolbox of amelioratives or the drug-hits of Hollywood imaginaries. Fourth, this other noon, or anti-noon, produces an inverse "twilight of the idols" in which climate comedy emerges, and in which a Thor-hammer is of less use than a quiet tuning fork or perhaps speculum. Does one write, today, whether one wishes to or not, before and for a somewhat different receptor organ than what we called, once, an evil ear or eye? It may be a bit corporeally provincial to speak of such isolated organs in the era of info-bytes and data harvesting, Facebook "subjectivities," and the "breakaway civilization" (Catherine Austin Fitts[13]) of the hyper-elite. That is, at the current stage, which is quite premature, in the spirit to be cultivated today, not of ethical outrage or advancing technocracies, not of environmental pragmatisms or sustainability rhetoric, but from the spirit of climate comedy—"as a recreation and spot of sunshine," and perhaps a new war?

In this imaginary, the term "twilight" is a bit jejune—holding on to a diurnal, natural, organic trope for what is not really an occurrence in such terms. Dawns, twilights, noons, midnights. The proper name "Anthropocene," of course, enters in its twilight, as a term with a very short half-life, engineered for distraction. And engineered for perpetuation of anthropos' universalist claims (now, in parting, to the entire human "era"). There remains a rather pragmatic question here, which can be posed with barely a straight face, assuming you cling to your inner anthropos still, gears whirring, recuperations pre-programmed, in the mode of resilience if no longer "sustainability" (or so the rhetoric). Why—again, straight face—in a decade in which societies were informed that a threat emerged to their collective existence and longevity, were they unable to come together in any fashion or form as one (let's not mention Paris, where now lobbyists will do the negotiations and where the numbers undercut the pretext). As we see today, in the summer of 2015, the "they" fragments into competing tribes of all sorts, vagrant nationalisms, closing off media and internet enclosures or, say, ecographies. China, for instance, would curtail "Western values" in its textbooks and devises a self-sufficient internet—all harassed by the background music of collapsing ecosystems, mega-drought, disappearing resources (water), and so on, with a list that is boring to tick off since it is endless in details and combinatoire? Why instead of a "universalist humanity" is one delivered to its competitive fragmentation, with independent memory systems and "facts" arising, the so-called West in the pack? The best one might say to those who, belatedly, want out of the imaginary Western contract, who smell how ill it has become or is, is to remind them how zombie bitten they are—did somebody say, Karl Marx?—and to recall the cynical decree, that what you hate most you will become. If there is or was an "Anthropocene" era, then there must have been this figure, anthropos, a Western concoction, hanging around and infiltrating or claiming the whole. But then, might cynicism here mean something entirely new, or disclose one of the many lapses applied to the term, a guard against that which does not accept the "we"?

2. Miss Lonelyhearts and the Deadpan—or, a Plague of Face

> *We must bear in mind that, fundamentally, there's no such thing as color; in fact, there's no such thing as face, because until the light hits it, it is nonexistent. After all, one of the first things I learned in the School of Art was that there is no such thing as a line; there's only the light and the shade. On my first day in school I did a drawing; it was quite a good drawing, but because I was drawing with lines, it was totally incorrect and the error was immediately pointed out to me.*
>
> Hitchcock to Truffaut [Truffaut 1963, 183]

In a recent essay in the *New York Times* Stephen Marche analyzes a new "epidemic of facelessness" in what we will call our *today*—in this case, it enters through a quirk on the periphery of our tele-sphere with a virulent logic [Marche 2015]. He is analyzing the effects of *trolling* on the internet, targeting web prey anonymously and with the unrestrained vengeful or defacing drive to violate. The link is not made to roiling para-movements on the seeming geo-political peripheries, like ISIS, with its Ebola-like spread and complete defacement of any residual aura or credibility of "the West" or hyper-modernities. This faceless attack mode penetrates the *real* despite its ambiguous status—including law courts and jailings for discovered miscreants. "Trolling" operates in the structural logics of terrorism. It can always be dissociated from by a legal individual who adapted a voice or persona and be called all "in play," as if in citation marks, a form of gaming virtually solicited by the *social media networks.* The dependency of such networks on photographs, or "Selfies," only conflates the dissociation, as persons become obsessed with representations of the face and movements (as if to reassure or leave evidence).

To situate today's "*epidemic* of facelessness"—the trope of illness pleads, at first, for a treatment—Marche presents a narrative in which *face* had been the foundation of an epoch of law, identity, and above all ethics (in the West). It was, or promised, a currency and economy of readabilities and debt, like, one might say, the Greek *autos* or, codified, *anthropos.* He indexes both Roman law's requirement to face the accuser and Emmanuel Levinas's claim of *face* as the premise of Western ethics and identification as foundational. Marche's claim is that *face* as such is dissolved, in the shadow-zones of *trolling,* into forms of faceless transaction that channel blind aggressions, including rape threats and take-downs of all sort, as baits for the trap of any response or engagement. (The target is advised not to respond, since it could unleash a swarm.) The *trolling* assaults are thus deemed "faceless" themselves, operating from invisibilities (or so thought). Such a *plague of facelessness* appears viral and predisposed within the unfolding of digital and other technologies.

It spreads, stain-like, from the obverse of a totalized frame. What is peculiar in Marche's account is that face is mourned as if a certain "we" were collectively leaving an entire techno-cognitive (and of course "social") epoch. "Face" of course returns us to the core of *anthropomorphism,* as we think of it, as if "face" were detached and sought in the world of things and forces (like Greek gods): it is bound to the so-called *Anthropocene* era inescapably and, as such, makes the latter dependent for its definition and ecocidal drive on, among other things, dominant cultural tropes or a more thorough critique of an epistemology of tropes (the purported focus, say, of de Man).[14] The first import of facelessness is dissonance, since the resuming of one's corporeal face or identity sees itself as disconnected from view, or that of surveillance:

> When the police come to the doors of the young men and women who send notes telling strangers that they want to rape them, they and their parents are almost always shocked, genuinely surprised that anyone would take what they said seriously, that anyone would take anything said online

> seriously. There is a vast dissonance between virtual communication and an actual police officer at the door. It is a dissonance we are all running up against more and more, the dissonance between the world of faces and the world without faces. And the world without faces is coming to dominate (Marche 2015, 2).

This may seem, at first, counter-intuitive, since in a world of screens, talking heads have become the norm, "identification" and "trust" manipulated (among else, corporate media ratings)—as have emoticons, and *emojis,* in which face is reduced to its bare markers or caricatures, distilled as an imagined (rather than expressed) *affect.* And yet, in itself, what "is" face, which we presume bespeaks presence, the mirror-stage, the mother's emotive bonding, but the organization (or individual subordination) of features, which now form an expressive or communicative, dissembling or hypothetical imaginary? A dog, cat, or horse, may or may not be given face; a machine, without facial musculature, cannot quite; and yet *figural* language—with which, together with emotion and ethics, *face* (or *figure,* in French) is confused—is defined as chains of substitutive logics. To "lose face" is confirmed as the nadir of social valorization and currency, becoming a non-citizen, a non-*person,* losing ontological purchase. "Face" in that sense becomes a commodity to trade. None of this has been lost on the televisual and digital sciences of face manipulation and technologies of binding, honed across all modern modes of "propaganda," screen manipulation, telemarketing, corporate mediacratics, and the "psychological" tooling of affects. In this sense, a culture of distraction dependent on screens feeds a steady diet of talking heads to perpetuate (and control) an economy of face that is now entirely simulated:

> the faceless communication social media creates, the linked distances between people, both provokes and mitigates the inherent capacity for monstrosity.... The challenge of our moment is that the face has been at the root of justice and ethics for 2,000 years....

This epidemic of facelessness has a moral component, or rather the ill or symptom betrays the latter's dissolution, and by inference argues for its restoration or at least transition to another iteration of that as a form of credit. This is followed by an account of Emmanuel Levinas's account of face as the originary bond of otherness:

> The face is the substance, not just the reflection, of the infinity of another person [according to Emmanuel Levinas]. And from the infinity of the face comes the sense of inevitable obligation, the possibility of discourse, the origin of the ethical impulse.... The connection between the face and ethical behavior is one of the exceedingly rare instances in which French phenomenology and contemporary neuroscience coincide in their conclusions. A 2009 study by Marco Iacoboni, a neuroscientist at the Ahmanson-Lovelace Brain Mapping Center at the University of California, Los Angeles, explained the connection: "Through imitation and mimicry, we are able to feel what other people feel. By being able to feel what other people feel, we are also able to respond compassionately to other people's emotional states." The face is the key to the sense of intersubjectivity, linking mimicry and empathy through mirror neurons — the brain mechanism that creates imitation even in nonhuman primates (Marche 2015, 2-3).

But before we affirm the narrative that face was originary and the key to ethics, discourse, self, and intersubjectivity—assuming these had ever arrived—it passes an odd loop, confirmed by neuroscience and exceeding (or preceding) "humanity," as "the brain mechanism" of "even nonhuman primates." This slippage of an irreducible humanness of the human to the primates might be winked at, did it not ascribe the entirety to a "mechanism" that exceeds human definition: mimicry or mimesis, or *imitation.* It would seem, here, that "mimesis" precedes face as such (were that to exist): some physical-affective-neural trope precedes the artefaction of face. That claim, too, is the product of a "brain mechanism" which spawns imitation, inscription, the bondage and bonding of a "we" (variously enforced). It precedes and is itself not particularly "human" at all.[15]

What is peculiar from the perspective of *climate change* is that, technically, the hegemony of face—of prosopopeia and diverse figural systems—represented the trajectory of an "Anthropocene" era that discloses itself as ecocidal and *autos*-cidal. From such an impasse, one might expect strategies of *defacement* to emerge. Yet rather than arriving as a critical agenda in a war, today, over inscriptions themselves, we find an "epidemic of facelessness" arrived in advance, in the province of digital networks—as if *trolling* secretly wanted not to disfigure but to penetrate and re-occupy face itself from the swarm of pixels and the time-space of the web. That is, the digital orders have simulated facelessness too preemptively, immunized it. And if that generates facelessness, in effect it always had, and what we call face concealed all along that it was a construct, a project, a technic. Of course, "face" is an artifice, selective, a composition of parts (sense organs, apertures, "lighting"), of shadow, of what is before-the-eye(s), pros-opopeia. What is disclosed, inversely, is not that there was face and that now we are entering a faceless era, unchartered, outside of "ethics," which Marche proposes, but that an epoch of "face" had been an artificial parenthesis—a technic, *anthropomorphizing*, itself "anthropomorphized" in citing its voice (or face)—within a normative state of facelessness, in which face was unreadable.

Perhaps, though, what emerges is not a snapshot of the new facelessness in a digital *telepolis*—for instance, the breakdown of contemporary ethics, even when replayed endlessly as about and "of the 'other.'"[16] Rather, what is complicated is not that an "epidemic of facelessness" should appear, but that it appears in advance, as if pre-emptively, of any act of critical *defacement* (allow me to adapt this term from de Man), that one might want to exercise, say, before an experience of *totalization* which the telepolis assumes?[17]

The impasse arrives in advance. Facelessness thus is ascribed first to the digital or technic platform *itself*, from which the trolling projectile is launched against the domain of "face" in the latter's parodic, celebrity-culture, selfie-traversed, emoji-expressed "social network" imaginary—itself tracked and perpetually channeled. Marche describes an epidemic of *facelessness* that precedes any critical strategy to deface, yet finds itself preceding the cultural contracts of *face* and *mimesis* jointly. Amidst the gargled idiocy of the individual moved to troll anonymously are elements launched against a totalizing culture of *face* (in the latter's own name), unreflectively linked across a chain of displacements to an event like ISIS—or, for that matter, to a Wall St. culture that has been "short-circuited" and is end-running itself in shocking manipulations aimed to extend and pretend. The troll is both a criminal before the community of face, to be cut out or arrested if tracked (as has occurred), and a hero in defying the artifice of a totalized and now artificed and neutralized "ethics" of face. Technically speaking, the troll's position is like that of the cynic, as one who does not subscribe to the "we" as a premise. The troll, exposing the artifice of "face" and striking at it, sustains the latter's economy by providing it with a negative gadfly anticipated as the circuitry's excess.

Marche's narrative has to do with the assumption that "face" is and had been a given, a historical core of the Western psycho-social and political trajectory—and specifically, ethics. To say the arc of the Anthropocene would be distinguished as *an epoch of face* would, nonetheless, call the latter into question. The non-trope or putative proper name of the "Anthropocene," as a speech act of the present, seems at once the last totalizing attempt to maintain a cognitive regime beyond its due date and an ideologeme without ideology. The non-trope of the *Anthropocene,* after all, arrives in and *as* its own posthumous "twilight," a cinematic flickering that may just be the terminus of a techno-archival regime (the *cin-anthropocene,* say), an Ozymandian inscription of mastery, which sees itself as the unfolding of *a specific human type, legatee of alphabeticism and the era of the Book*—the Greek *anthropos* (for the codifying Aristotle: Greek, male, and a possessor of reason, or *logos*).

Marche is aware that this recollected institution of face persists in and as figurative language, or *figure* as such: this giving of human face to the world of things is what, typically, is associated with *anthropomorphism. It* is the latter that we assume we recognize and curtail in ourselves, at least, as sophisticates. *It* is what has led through attempts to pivot outside of a tired "anthropocentrism" by identifying with allo-human others or othernesses—as if no one wants to be tagged as *human all too human. It* is the supposed function or constitutive apparatus of what was called "language" (or consciousness, or both) that welds and harnesses a mobile figure of humanity. *It* has led through ventriloquizing dogs and animals, stones, plants, "objects" and *things*—the *inanimate,* in short, even while the spraying of otherness as a way of cracking the spell is swathed in literary assertions of *affect,* bathing again in the claim of restored instinct and a rather pathetic assertion of emoting (if not pathos). *It* is what we assume to be naive: a technology of perception in predatory anthropo-narcissism. It gives voice and seeming animation to things (and animemes). No one, however, seems to have stopped to define, articulate, or analyze *anthropomorphism* as a reflex or artifice, trope or technic. That is, no one stops to ask, it seems, why a *figure* that founds the "Anthropocene" itself does so by naming the role of cognitive rhetoric in the acceleration of "climate change"—that is, among the mass spells, perceptual disconnects, and digital pacifications of a populace in virtual climate denialism despite "debates," accords, corporate plans for geo-engineering, and a rhetoric of *adaptability* and *resilience*, inevitable as "tipping points" pass. A politics of managed extinction, a segregating out of disposable territories, geographies, populations, adheres to this event. *Today*, the implied criminality of facelessness rises at the same time that facial recognition technologies consolidate data-bases.

Marche bemoans a loss or disappearance of face, and with it a model of ethical "man," for the replacement of which there are *no psycho-political models*. This would rhyme with Timothy Morton's overt indexing of cognitive and moral weakness attending an era of *hyperobjects*—distinguished by being inaccessible to perception and programmed senses, as if that disconnect echoed in Wall St.'s mafiatization, the corporate capture of "democracy," and the hollow disconnects of "Enlightenment" (or utopist) rhetoric.[18] Yet one may toggle between the macro and the micro here, between the hyper-object and what, too, begs attention and conferred agency, the errant and unreadable hypo-scripts that subtend and traverse the telepolis' screens. Marche thus overlooks that the artifice of a defacing avatar or web-persona mimes "face" to gain the discourse community's entry—and calls out the artificed milieu of the similarly web-presented success of the prey target: the *troller* would participate in and conveniently dissimulate that the narrative arc of a community of face was always artificed, and was always one organizing technic or con among others.

But if a *neural-mnemonic-tele-political regime* resulted and brought about an ecocidal hyper-consumption, one that appears totalized at the very moment its sheer artifice and techno-dependence is disclosed—there had always been an invisible *Facebook*. Moreover, would this not confirm less an epidemic of facelessness, than the unwinding of an era, perhaps something like the "Anthropocene," which itself had culminated in a *plague of face*? Might one not think of "face" along the lines of facial recognition software, in which points align and assemble in unique relational mappings to trigger "identification," something that may be learned and counter-feited. Might one think then of an era of "face" as one in which that software had been programmed in, and no doubt tortured into the mnemonic commons which could revoke it in turn, and the co-ordinates assemble like a second skin over the musculature, like a scaly encasement? Hence the compensatory proliferation of selfies—each a doubling-down defiance of the extinction advanced by it. (The promiscuous "selfie" today moves the game-board forward one or two innocuous notches, knowing that every photograph poses before the eye that might receive it after the *catastrophe* or erasure—and every *selfie* squeals with that discrete logic.)

The logic of *the promise* that sustained the credit cycle of Western discourse and its monotheist technic is in effect vaporized, or exposed as hypothetical, by twenty-first century climate change. The latter implies that collective *futures* may be structurally revoked by extinction events already set in motion irreversibly. And if the promise of *the promise* were *post facto* disclosed as void, together with any claim to futures, it would necessarily be voided retroactively too—altering every text in the archive that depended on it to be "properly" interpreted. While trolling from a position of facelessness constitutes an anonymous aggression seeking to enter the picture, in order to be felt and to assume "face," *face* is exposed as a commodity to inflate or trade and an artefaction preceded by a phase of mimetic imprinting. An installed accord, entropic, it maintains referents or generates them—as if cinematically:

> The connection goes the other way, too. Inability to see a face is, in the most direct way, inability to recognize shared humanity with another. In a metastudy of antisocial populations, the inability to sense the emotions on other people's faces was a key correlation. There is "a consistent, robust link between antisocial behavior and impaired recognition of fearful facial affect. Relative to comparison groups, antisocial populations showed significant impairments in recognizing fearful, sad

> and surprised expressions." A recent study in *the Journal of Vision* showed that babies between the ages of 4 months and 6 months recognized human faces at the same level as grown adults, an ability which they did not possess for other objects. Without a face, the self can form only with the rejection of all otherness, with a generalized, all-purpose contempt — a contempt that is so vacuous because it is so vague, and so ferocious because it is so vacuous. A world stripped of faces is a world stripped, not merely of ethics, but of the biological and cultural foundations of ethics.... The spirit of facelessness is coming to define the 21st century. Facelessness is not a trend; it is a social phase we are entering that we have not yet figured out how to navigate (Marche 2015, 3).

But what if the posited and enduring "we" were itself an effect of this regime, including how "we" think or iconicize "the" social or "the" political or have all along—the "anthropocene" signature and default ill in advance, willing to read everything but its Ponzi-stretched and captured indexal "economy"? The "we," after all, is non-existent technically, artefacted within an anthropomorphic assertion and a by definition ferociously *exclusionary* regime—the *we* of a privileged circle, specialist language, gentlemen's accord, shared bet, criminality, or usurpation. This renders any figural hypothesis of *humanity*, of a single or shared humanity of hominids, as impossible as it is to locate the authority of the "we," which is a rhetorical technic promising to dissolve (and re-stamp) individuation. Such a "we" is a mutation of the same "I," or *Autos*, the *it* or inimical "they" that opened the Greek grammatical adventure. So while this same "we" supposedly figures out "how to navigate" the "new" orders of facelessness, how to avoid its aporia and steer around its black holes, at least in order to perpetuate this "we" (we, purported humans), or sustain the assumption that the *social polis*—like the "Anthropocene"—is a premise immune to any suspicion itself, it is unsurprising that corporations have moved in to appropriate not only the legal role of "person," though incorporeal, but then, pushing on, claim the domains of "human" affect, spiritual prerogatives, anti-abortion stances, religious freedom, care of the unborn human (with, as it were, a straight face). Yet is this non-existent and somewhat corporatized "we," in fact, or ever strictly, as is asked of Lord Jim, "one of us"?

One would think this new zone of pre-emptive auto-defacement were a threat—but it catalogs a mutation that is "not yet figured out," even if it conjures an outside or precession of *figuration* itself. The promise's ghost lingers in a robo-stutter pointing to the next one to come along, who will figure it out, an open deferral. Needless to say, this mimic zone mimics itself in the digital short-circuiting that *produces nothing but faces*:

> Google recently reported that on Android alone, which has more than a billion active users, people take 93 million selfies a day. The selfie has become not a single act but a continuous process of self-portraiture. On the phones that are so much of our lives, no individual self-image is adequate; instead a rapid progression of self-images mimics the changeability and the variety of real human presence.

Marche cannot yield to a nostalgia of "face," which is irrecoverably dissolved across digital accelerations (and homogenizations). One is now in the imitation of faces, or tags like emoticons. Yet he cannot allow that this welding of face-granting "intersubjectivity" to what might be termed *Anthropocene 101* "ethics," as viewed in the rearview mirror, may itself have been but one random organization among others, a time window like the post-war Western "middle class," or the little ice age whose passing sprouted Enlightenment memes. Since if *face* were always an artifact produced by machinal and mimetic triggers, it was always also one technic among others. In this case, it is one welded to a certain regime by which identification, referentiality, and totemic recognition can be provided an economy. *It*, of course, is de facto a front for con games and manipulations, one that makes Melville's *Confidence-Man* appear shy.

Would the so-called "Anthropocene" or its simultaneous twilight, today, coming out of the model of the Greek *anthropos* and Roman law in the West, not appear in a sense as the time of faces, as Marche implies? This would seem to resonate with the ritual manner in which something like "anthropomorphism" is casually referenced as a lamentable naiveté as if gone beyond—without one ever stopping to define or analyze the term (anthropomorphism) or what it may do if it all rested on a single trope or metaphoric regime, assuming that is what *anthropomorphism* connotes? A technical problem with "face" is that in reality it depends alone on light hitting it, and it depends on an eye seeing that *and* as if being seen by it—as Hitchcock tells Truffaut: "There is no such thing as face." On the screen and in the world in fact there is "no face," just the way the shadows and light coalesce. That is how *prosopopeia* can be transposed: defined as the *poeisis* or making of what is before the eyes, as face, the giving or emergence of voice in the inanimate, the coalescence of the sense organs of the head (ears, nose, eyes, mouth) in one cognitive place or center, and finally their capture, today, by facial and retinal recognition security systems.

Thus any fetishization of "a facelessness epidemic" must be questioned for its inverse or hypo-defacing logic—quite aside from the nostalgia for "face" that is tracked in a sort of double-chase, the hunter stalked. It is a mimetological war that is witnessed, since the mimetic accord both engendered face (mimicry) and is sustained by it (identification, posters, Leaders, selfies, televisual screens, talking heads). Are we sure "face" has not been a problem—at least, if taken as a mimetic regime, which is to say, one that is technically replicable and subject to *capture* (tele-marketing of all kinds, now "pre-crime")? Is it not, at least, indissociable from a certain, obvious, "anthropomorphism" or even the Anthropocene as a hashtag for ecocidal disappearance? Is or was "it," one organization of visibilities among others, itself politically consolidated or *used*—as when *arche-cinema*, in Stiegler's insight, both enables archival communities and perceptual grids yet allows stereotypical grammars to institute themselves, such as appear totalized today, at the apparent "death" of cinema and its migration from screens to interactive interfaces and neural management?[19]

Facelessness, a turning against the regime of face, is pre-emptively claimed and programmed by and within digital streams *in advance* of any cognitive or aesthetico-political turn. Let us posit, for a moment, a system, call it a corporate system today, from which decisions are disseminated and to which all hyper-technologies appear back-looped if not subordinate. It pre-empts and takes the strategy of defacement away from any position of resistance, and in fact perpetuates, for now, the ghost ideology, which is to say the aesthetic ideation-ography of "face," in a streamed, managed, and anaesthetized mode. Every *selfie* is a zombie lunge for the blood of a self-penetration—which is why I suggest we look, on occasion, to how a "selfie" of the Anthropocene itself might be posited, marking the lens' and cinema's technical agency in what unfolds today (what one might name, if at all, the *twilight of the cin-anthropocene*). Thus the "epidemic" Marche ascribes the position of facelessness to: as to the technologies that irradiate or rage through the trolling troll, say, from its shadows, unseen and wielding the "Ring of Gyges" effects, invisible itself (it thinks), dissociated into some irretrievable dark side of gaming itself (from whose ranks drone pilots are drawn). The magical "troll" is neutralized as anything more, contained as what presents a symptom of a wider "present" (hence Marche's generalization from a peripheral harassment we are advised to ignore if it occurs to *you*). Of course, the import of such a staged logic of defacement in a broader sense links up with the always black-faced ISIS horde of video beheadings or the resolute refusal of any metaphor by Ebola (or the tendency to link both as media memes).[20] That is, say, to the virulent and inhuman agent, borderless and hybrid, capturing the technologies of cell formation and dissolving these as a mobile, Borg-like imaginary shorn of face, trope, metaphor, recognition. Outside *mimetic* orders, the videos of ISIS beheadings crystallize the rebuke and ridicule of mimesis in the West (which Islam also rejects): only the head of a Western face will do, held up on video to the West, cut off by a British voiced rapper, an auto-beheading that lures, say, Obama into the mediatric necessity of another losing war he wanted at all costs to avoid (no war or response, of course, for hundreds of thousands of murdered, raped, crucified or beheaded North Africans, Arabs, Christians).

Is there reason, then, to ask if a destroying era of *anthropos*, such as it is or was, ran on more or less tropological regimes—heliotropic, humanualist, organicist—in which an economy of face and the installation of a mimetic "contract" of reference production would be directly wired to the unfolding logics of planetary ecocide?

The problem in Marche's account is the obverse of what one might expect. It suggests that a disappearance of face crashes a pre-digital regime that separates and lingers on, artificially streamed, in pointless selfies or televisual screens (where talking heads abound, re-enforcing identification). And it is not clear what accord will come next, particularly if it is shorn of the means of harnessing care, empathy, "ethics," credence. Here, though, is the problem: such an entire regime trading on face and identification, subjectals and "inter-subjective" contracts, was itself the symptom of an ecocidal, hyper-industrial trajectory—passing into the autonomous *telepolis* whose dark economy the supposed *ill* of an "epidemic of facelessness" confers. It is a dark economy in the same way, today, as we find the international cartels moving drugs and trafficking humans globally intertwined with international banking, which would collapse without its illicit liquidity (HSBC). It is the production, marketing, and assertion of *face*, together with a legacy of cognitive and figural conventions, that is interwoven with ecocidal trajectories—and would need, like "anthropomorphism," to be interrupted, undone, disfigured, defaced, to get outside of the spell or trance. Potentially.

But here is the evil twist that is implied. When there occurs a societal totalization of a system of tropes (a "culture"), which is also to say of perceptual regimes, so that nothing appears outside of it to those caught up in it, the violence of defacement converts into a second-tier strategy. (I will take up in a moment what that entails using de Man as a foil or catalyst—and, perhaps, Nietzsche's claim to an "evil ear" in hammering the idols of what he presents as a noon moment, which would appear deflated and inverted, perhaps, before the horizons of "climate change" today.)

Marche does not make the connection to the era of "climate change," to the internet of *things*, to the wars of last man culture waged on the peripheries or dark sites (societal, economic, geographic, digital), nor move outside of a contemporary dilemma redoubled by digital media's totalization. He does not, for instance, probe where this now disappearing era of *face* may be regarded as itself artificial, deceived, cinematized, narcissistically destructively, not specular but a sort of installed or inserted technic. So-called primitive "tribes" that practiced animist modes of trans-spiriting make a point of marking the face and projecting masks of dream and power: these avoid the problem of "face." One can always "put on one's face," as projected. "Face" would be the devising of two stabilizing traits: like today's facial recognition technologies, it would enforce security, define the group, be the guarantor of identification in the theatrics of leadership and violence. It would serve a purpose, and be commodifiable, as body modification is aware today. It would appear at least coincident with, as Marche notes, the idea of facing one's accuser, and the face-to-face encounter as a facade. "Facelessness" coincides with a loss of what Benjamin's calls *aura* and the personification of the inanimate. The "trolls" Marche cites are, accordingly, monsters: caught within a leftover economy of face they cannot escape, as it is streamed and regulated, they mime the position of the prehistorial and premimetic, rehearsing predatory raids on the screen icons and reserves of this currency (*Facebook*, say). Yet they do so from a digital position that determines in advance a techno-facelessness over which it has no traction or decision. The pathetic raids on the spectral kingdom of face (celebrity culture, twitter feeds) practiced on stray individuals as a sort of digital rape or kill (suicide epidemics for teens), is itself a trap, a toxic economy of maintenance in which a simulant and streamed economy of face is maintained as sport and distraction, with the effect not of a loss of ethics and address but of the regurgitation of modes of identification, mimetic coding, the pursuit of "aura," and ethicist rhetorics bent to shore up a closed system or epoch.

3. Mourning becomes electric—or, Greek words bearing gifts...

For all of the collective decision to occlude the question of language and climate change, the last thing negotiators, scientists, and contending activists need give attention to, the least real or relevant to extinction processes and the undoing of a tentative terrestrial balance or calm, one still hears routine finger pointing at *anthropomorphism*. It is a term supposed to spray human projections onto all *things* of the world, a trope of tropes so banal as to garner no suspicion itself. Those who take it for granted never stop to ask what "anthropomorphism" is, names, or projects—nor, even, why it is deemed bad. The same happens to the human speaker: that the grammatical "I" too is anthropomorphized in and by social language, given face, aura, person(hood)ification on credit, so that one's voice or persona or self is credited. Given the current stakes on the table, which involve the course of species life going forward aeons, one might think this would be flagged somewhere? So far, however, that is quite timid. Thus a certain *we* finds itself passing ecocidal tipping points, without fluster, and having suppressed the question of language's relation to climate change, biomutation, and extinction events.

This question is uniquely if not solely raised where one might least look for any input whatsoever, in Paul de Man's apparently very *last* and posthumously published essay, "Anthropomorphism and the Lyric" (De Man 1984). It is a murky piece, since it leads not from the random zones of rhetorical blinds out toward the worlding of worlds, nor to "things." *Climate change* throttles wholly in reverse, here, back into the vortices and black holes of what generates perceptual regimes and tropological displacements. One can easily overlook the connection to today, as the address seems minor, returning to a line of Nietzsche's "On Truth and Lie in an Extra-Moral Sense" (Nietzsche 2009), a touchstone text for de Man's approach to rhetoric, tropes, cognition, reference, and so on. Of course—if one needs reminding—the template of Nietzsche, like that of Machiavelli or, read well, Plato, understands language as ordering itself out of random fictions, organizing lies, strategic dissemblance, that is to say, *tropes*. If there would be an ecocidal trajectory emerging from "modernist" and hyper-industrial legacies, as now is clear, it would emanate from the cognitive settings and tropological regimes that informed that.

Nietzsche's essay fragment invokes (and disposes of) something like an "Anthropocene" moment in its first two lines: that is its premise or "thought experiment," in today's captions. It drifts into the question of a cognitive ordering of the perceptual world from tropes, but begins with two other, more decisive questions: the first is the situated role of "intellect" in his crafting of the world, which is regarded as a tool or function, separated out from *anthropos* as such, whose primary tool of self-preservation is deception; and the second, the epistemology of tropes, which can themselves only be viewed as in perpetual excess of "moral or ethical sense" (human on human calculi). I cite a medley, which reads like a toss-away pin-prick on any "Anthropocene" *avant la lettre*:

> In some remote corner of the universe poured out into countless flickering solar systems there was once a star on which some clever animals invented knowledge. It was the most arrogant and most untruthful minute of 'world history'; but still only a minute. When nature had drawn a few breaths the star solidified and the clever animals died.—One could

invent such a fable, but one would still not have sufficiently illustrated how pathetic, how shadowy and volatile, how useless and arbitrary the human intellect seems within nature. There were eternities in which it did not exist; and when it is gone nothing will have happened. For this intellect has no further mission leading beyond human life. It is human and only its owner and creator treats it as solemnly as if the hinges of the world turned on it. But if we could communicate with a gnat we would hear that it swims through the air with the same solemnity and also feels as if the flying centre of this world were within it. There is nothing so reprehensible or low in nature that it would not immediately be inflated like a balloon by a small breath of that power of knowledge; and just as every porter wants to have his admirer, so the proudest of men, the philosopher, believes that the eyes of the universe are trained on his actions and thoughts like telescopes from all sides.

It is remarkable that this is accomplished by the intellect, which is at best assigned as an aid to the most unfortunate, most delicate and most transitory beings in order to detain them for a minute in life, from which they would otherwise, without this extra gift, have every reason to escape as fast as Lessing's son. Thus the arrogance connected with knowledge and sensation, covering the eyes and senses of men with blinding mists, deceives them about the value of life by carrying within it the most flat tering evaluation of knowledge itself. Its most universal effect is decep tion, but even its most particular effects carry something of the same character about them.

The intellect, as a means of preserving the individual, unfolds its main powers in dissimulation; for dissimulation is the means by which the weaker, less robust individuals survive, having been denied the ability to fight for their existence with horns or sharp predator teeth. In man this art of dissimulation reaches its peak: among men deception, flattery, lying and cheating, backbiting, posturing, living in borrowed splendour, wearing a mask, hiding behind convention, play-acting

> in front of others and oneself, in short, constantly fluttering around the single flame of vanity, is so much the rule and law that there is hardly anything more incomprehensible than how an honest and pure drive for truth could have arisen among them (Nietzsche 2009, 253-54).

What has this to do with the cataclysms of melting ice sheets, mega-droughts, or the inevitable unfolding of mass extinction events, supposed to be so much an issue of science, carbon counts, feedback loops, a hyper-financialized economics of extinction, and so on, as to make any such thought *de trop*? Or: precisely the opposite. In any case, de Man's last essay draft examines "anthropomorphism" as it appears in a single part of a sentence in Nietzsche's text, and it then moves through a relational reading of two Baudelaire poems—"Correspondances" and "Obsession." Who would expect a logic to emerge that would be of relevance to twenty-first century aporia of biomorphic mutation, accelerated ecocide, or "climate change"? Have any logical aporia in this short archival history (a couple of millennia or so) ever *not* included this as an implication? I want first to conjure a very different context for approaching, here, why any discourse that puts into question the techno-genesis of today's "we" requires a more complex alertness to the hazards of reading after *tipping points* are past.

Nonetheless, as we know, *anthropos* likes to hedge his bets. No one wants to give "anthropomorphism" up altogether, perhaps, nor can—unless there were a more primary narcissism still that precedes that, before "face" is posited or seemingly contracted, self-archived and tagged as a debt, a promise, an assurance. And even then, no one wants to give up certain things: their energy expenditure, their smartphones, their investments, their credit with diverse mafias, or the perks of "personhood" held hostage to a system of privileged exchange (credit cards)—yet exclusion from which triggers new zones of disposability.

One has slipped far from the Nietzschean question that disbands "the human" as a cognitive or epistemographic primitive to be hyperbolically recast to survive itself. As if in caricature, literalized and stripped of import, this prospect is moved to the "post-human" column conveniently. The latter has been reduced to mean merely how to outwit his mortality, replace his mortal body, download individual memory, in short, make the present of *Last Man* culture frozen and perpetual against time-space, the organic, corruption. "Anthropomorphism" is cast about as something we can counter or suspend, or point to occurring elsewhere, a tautology seldom analyzed or defined so much as just blankly assumed. And what it assumes is a projection of human qualities onto things and entities that subject them to our extractivism and economies: they assume that this given, *"we anthropoi," is a given* and can be exported into the inorganic and merely animate, since it is "we" doing so. It is, or seems at least to be, a *trope*. As such one might imagine, if nothing else were working (it isn't), that the mechanics of tropes would be one of many zones explored by sophisticated interrogators—yet, on the contrary, this key premise is avoided, as if some unpleasant secret lay within its premise.

In advance of the moment when everyone had had enough with the "nothing outside the text" conundrum, including a Derrida frustrated by its polemical capture, de Man doubled down. It remains interesting that a Quentin Meillassoux (2010a; 2010b; 2012) would pivot on a break with "correlationism" (in-bred cousin of mimeticism) to supposedly occlude (again) deconstruction and language in order to access the contingent, whereas de Man—who is not to be taken as deconstruction—dissolves *back* into a "materiality of inscription" that departs from the *non-relation* of what was only nominally called "language" any longer (since that could not be dissociated from cognition, mnemo-technics, perceptual regimes, events), the *dissociation* of rhetoric and grammar, of trope and violation.[21] Timothy Morton one-ups the relapse back from Meillassoux into "objects," *OOO*, by invoking a *hyperobject* which is not present to the senses, and lays bare a weakness and cowardice that adhere to its time among men. The prison-house of generated "reference" and referents coalesces into more general trances of hyper-industrial post-global tele-mediacracies: infobytes condense, memes iterate, puffs of affect go viral, mnemonic entertainments barrel-bomb media. The heir to such totalization echoes the capture of mnemonics by telemarketing techniques—and by its occlusion of "climate change," its inability to respond to or see what is before its eyes, technically. Does an *epistemological* critique of tropes hum in a particularly fatal way in the era of climate change when, as de Man discerns, figuration in its entiretly can appear caught in a short-circuited vortex of substitutive chains, capturable or subject to (now digital) totalizations?

And yet: "anthropomorphism" is the *topos* of de Man's posthumous and perhaps "last" essay—a curious exit, or opening. "Anthropomorphism and the Lyric" is a demure title, until one finds the two are not the same, or even what they appear to name: that is, that "lyric" is a by-name for a tropological condition of human positing and that "anthropomorphism" is unlike any trope and is technically not one at all. It is up to something. Might one engender an entire critique and overpass of the "anthropocene" by playing out remarks on this one topic? No doubt, today is marked by a totalization of the screen, tropological bots that guide, harvest, and would direct perceptual and mnemonic circuits. Were this the rise of a corporatocracene, it might explain why the term "anthropocene" was allowed to circulate as a tempting but self-referential break to ponder, before submitting the term to corporate aims, bolstering geo-engineering and "adaptation" rhetorics to come. It would be vanity candy, laced with toxins.

I'll re-iterate the "de Man" that interests me here—a twentieth-century post-anthropocene thought that may or may not point to what has been occluded from the elephant walk of "climate" discourse and politics for the past decade largely. One of the ironies of the "present" is the degree to which the very impulses that drove the restoration of new historicisms and progressive "social justice" imperatives lapsed back to mimetic protocols and utopist rhetorics wired to the same ecocidal engines as the protocols of the right. Was *mimeticism* itself a program that once installed, say, by an *arche*-cinematic accident, guaranteed not only extractivism in advance of any formation of Capital—including today's mafia or corporate "capitalism" of gaming systems and financial weaponization—but also ecocide as its premise, and was *it* further captured, hacked, and subjugated by the equivalent of tele-advertising bots? Thus when de Man addresses "anthropomorphism" by returning to Nietzsche's "On Truth and Lie in an Extra-Moral Sense" one might take note at the manner in a which a place-holder term with no definition, *anthropomorphism,* is lazily cast about and even castigated as a given, often by its most cunning agents:

> But 'anthropomorphism' is not just a trope but an identification on the level of substance. It takes one entity for another and thus implies the constitution of specific entities prior to their confusion, the taking of something for something else that can then be assumed to be *given*. It is no longer a proposition but a proper name.... (De Man 1984, 241).

Just when we might have hoped the figure of personification or giving face were the navel of utterance, out of which a posited self or "I" is shaped—we are interrupted, cut off. Even personification or *prosopopeia* is, already in effect, a *defacement*: "The lyric is not a genre, but one name among several to designate the *defensive motion of understanding*, the possibility of a future hermeneutics." That is, when we try to identify in a specular fashion the moment of projection, by which the anthropomorphism of the earth will be read and the reserve of the human face or voice registered, we find that itself is already "defensive," a secondary reaction, and this to preserve the "possibility of a future hermeneutics." So, we learn that even *face* is a citational artifice in the service of keeping a bourse or exchange open; it is secondary, socialized, monetized, (in) corporate as a derivative bet on a "future hermeneutics" (one which it would practice, *as if*, in the meantime: less as promise than as promissory note). Defacement does not have anything to do with the hacking away of a monument or statue in place already. It hovers as an antonym, a phantom or echo-positionality for what remains before the contract of "face" at all (let's call it a positionality, for the moment, since this can be cultivated I assure you). One cannot routinely speak of a blackness in the absence of light altogether, or before the latter's techno-genesis. So, one cannot speak of face before a "defacement" that can only be named as a negation by the trope that had not yet been confirmed as in place to say nothing of formalized, assumed, or socialized.

Any supposed problem with *anthropomorphism* might stem from the fact that, while *passing* as one of the gang, it is not a trope or rhetorical figure (unlike metaphor and metonymy). It is not even a word, but rather a "proper name" for an effect that reverse-narrates its import. Rather than project onto the entities of the world "anthropo-narcissism," infusing things and beings with its own traits or affects or sometimes voice—speaking animals, cartoons, faces on mountains, diagrams—we witness a three-card monte that is not very surprising. It is anthropomorphism that, inversely, retro-projects that there was an entity or *anthropos* to begin with: it "thus implies the constitution of specific entities prior to their confusion, the taking of something for something else that can then be assumed to be *given*." Anthropomorphism is a "proper name" for an entity that was not there definitionally, and may never be the referent of the language it projects to anthropomorphize the world (as it then assumes). De Man turns this difference into one not only between tropes—which at one point can appear as being all in essence *the same*—and something impossible to personify or anthropomorphize in any way, inscriptions unreadable and without access to sense, to perception, to phenomenality, or to memory (since they generate each as software code). When de Man pairs Baudelaire's "*Correspondances*" with the latter's very different "*Obsession*," he sets the two poems against one another, exposing the non-relation between them and the mode by which *inscriptions* appear to spawn the rhetorical figures that counter-generate and disfigure them. Of "*Correspondances*" he says finally: "there is no term available to tell us what '*Correspondances*' might be. All we know is that it is, emphatically, *not* a lyric. Yet it, and it alone, contains, implies, produces, generates, permits (or whatever aberrant verbal metaphor one wishes to choose) the entire possibility of the lyric." A writing that precedes figuration yet "implies, produces, generates, permits" and so on the latter's arrival is possible to narrate as preceding the latter, temporally, but that is irrelevant ("'Obsession' might have been written before '*Correspondances*,' it would alter nothing.") There is a non-site which antecedes trope and its totalizations in situations, ideologies, epochal and geographical communities—what might be called in *The Birth of Tragedy* a "primal dissonance," or in Benjamin the contemporaneity of the "prehistorical," or in Derrida *khora*, or in Stiegler *arche*-cinema, or on Serres's work on Rome the black origin or 'black box', a zone of cow paths before mimetic contracts have gelled or been enforced. De Man sometimes used a boringly dislocated term, *literariness*, to invoke what antecedes the participatory systems of exchange.

The import of identifying the stupid or fiat nature of "anthropomorphism" is not that it takes the projection of man as a default condition. It is not even a pre-originary theft—that is, one that is outside of the artefaction of "origin." The "I" or so-called subject (this twentieth-century bane-term) is only conjured by whatever violation has already occurred. In giving the proper name "anthropomorphism," it assumes and pretends to conjure the *anthropos* as a retro-hologram to confirm itself. This account is opened as a Ponzi scheme sustained by an imaginary of fiat credit given (by itself) to the promissory. The same error haunts usages like "anthropogenic climate change," where a distinction is harbored between its public sense, that man's actions cause this and that destruction on all sides, and a reflective one, in which whatever "man" was and effects, "he" does not have control of the processes that produce him, too, as *homo climactus* or *homo ecocidus*—something quite distinct from critical fantasies of ecological man who, we will now be scolded, was all along a front for a *corporate drive*. The problem here may be that *anthropomorphism* is disfigured by its own proper name, removed from trope, word, or defined event. Hence its collusion with a pop term like "Anthropocene" today. While the latter is circulated by geographers as descriptive, even literalized to the degree they bicker over timing (the latest: nuclear fission, Hiroshima, which will surely be in the strata as a marker!), the conundrum adheres to the speech act that asserts the name at all. That involves "naming" oneself as and with an entire geo-destructive epoch (with a certain pride), and ostensibly in a manner that implies and can only be confirmed (or in fact read) after the fact, with a back-glance, with other eyes altogether—a "possible future hermeneutic" which one must go extinct to test.

"Face" was never about presence, but rather seduction.[22] Even the seduction performed by making facial expressions is transacted through learned responses, manipulations, "acting as," from the start—the infant calibrating the mother's response to what is mimed and brought. *Mourning* itself becomes a structural device, much as it is triggered in advance by the absence or loss of whatever is named or photographed—activities that are not "descriptive" but appropriative, archiving and negating in advance. It is precisely in the moments that we *feel* ourselves as most humanly human, most confabulated by affect, such as mourning, that we are most deluded—as our robo-programming by tele-marketing and anaesthetizing or self-cancelling Orwellian memes make too clear. It is the "anthropos" and its affects that are neither authentic nor originary.

It would in this narrative cease its fiction of an exponentially consuming *oikos* or *polis* or, detached from organic life or beings, head or body, a "corporation," if it unhooked this circuit or "short-circuiting" (Stiegler). It appears glued together, so to speak, by some master trope, as it seems, tied to *prosopopeia*. That appears to be *mourning*—what itself seems triggered by writing, photographs, and cinema as a quest for a recollected loss which was not any more specific than the trauma it associates with that. Here it is "mourning" that, in effect, dissembles the import of there not even being loss to begin with. What is taken away, and what only is experienced as a lost capital, is the dependency on a phantom mediation, a relapse or recuperation, a mimetic or historicist attestation that is substituted en masse for inscriptions. Thus, its other:

> True mourning is less deluded. The most it can do is allow for non-comprehension and enumerate non-anthropomorphic, non-elegiac, non-celebratory, non-lyrical, non-poetic, that is to say, prosaic, or, better, historical modes of language power (262).

Now, for a "final" essay, it is a sort of inchoate but precise last stab (inchoate, if one finds in Baudelaire, if not Homer, the zombie army of totalized tropes already, but again, it might as well be Homer). It involves a peculiar casting off of mourning, which remains a prevalent rhetoric of castigation, now, for the future: can one mourn "futures" despoiled and cutting off generations? One may argue that this refusal of mourning, too, was implied by Benjamin, and steered away from, not convincingly, by Derrida, who would run up the ladder to "absolute mourning," for operatic presti*digitation,* as he would with "absolute hospitality" (a blind totalization). What here is "*true* mourning" if it does not mourn, any more than one *must* mourn as such a human extinction logic, "anthropogenic," or in the name of what disappearance or "life"? This "true mourning" is defined against a representative roll call of trances, affects, poeticizations, affirmations, restorings of loss (mourning), integration. Here all tropes gather on one side, whereas the other side appears not to be without trope exactly—but as something among them but not of them, and lethal, making of them an "army (*Heer*) of tropes."[23] This draws any putative logics of defacement, today, into focus. It is a logic that, famously and early, deconstructs "deconstruction" *as such*.

De Man halts. He halts before an array of terms that are not figures finally, but more like opaque things implying cracks and intrusions in a nominal order based, in the end, on substitutive logics (all tropes essentially). These are intrusions, nonetheless, of a language of power that deform networks of sense and refuse metaphor, face, or *prosopopeia*—that is, the assumed correlation of figurative and affective mechanics. It is, essentially, in advance but also to no great surprise, a defacement of the Anthropocene as the by-product of a running anthropomorphic hum and machinal premise that we do not stop to inspect—and, specifically, the homunculus of the "anthropos" as Aristotelian devised it (not as subject). Since the "anthropos" is posited and then perpetuated as a mnemonic hologram retro-projected from the acts of linguistic appropriation and expropriation that are called, inversely, anthropomorphisms, one is positioned differently. There is no "human" to overcome or be more sophisticated than (be post-human in relation to), since *arche*-biosemiosis, and hence "consciousness," is not specific to hominids nor, as Benjamin strictly observes of language, "human"; there never was anything like a metaphysics of presence understood as an epochal effect or historical decision once taken—but only the perpetual simulation of that brought about by a hermeneutic reflex and "relapse" in the perpetual if artificed present. Yet one can add, here, as an aside (and in agreement with Claire Colebrook), that de Man did not go far enough, focusing on destroying the routes of retreat, sabotaging the recuperative reflexes one can now see as ecocidal in import and extractivist in "origin."[24]

Inscription as a non-word traverses stone, strikes, engravings, proliferating inked archives, robo-marketing, yet it belongs to non-living orders. There is nothing particularly uncanny about the prosopopeia of a stone, which speaks from the non-living or, for human consciousness, the dead, since living death is the condition of linguistic consciousness, as of "bio"-semiosis (semio-animation, bio-mimetologies). Similarly, today's tele-citizens cannot apprehend climate change or extinction events as other than another movie, an "as if," entirely disconnected from the mutations it experiences and receives data-streams on now daily, as part of the occlusion. De Man's use of *unreadability* is not to be heard as (or applied to) the undecidability of mutually exclusive interpretations. Rather, this unreadability becomes inflected from the perspective of "climate change" and the post-Anthropocene, much as one regards the Nazca lines as unreadable. It accords with the unreadability of a writing effect that, looked back upon, delivered its bearer ("anthropos," if he existed) to a vortex of extinction events, ecocidal accelerations, and the erasure of "life as we *knew* it." Here is where I begin to have my second thoughts on de Man's "use" for thinking critical climate change, at least at *this* juncture—but first, allow me to expand this rift or division. I will give a hint, nonetheless, as to one place this leads, I have found, which seems to be too embarrassing to contemplate usefully (embarrassing, that is, for a planetary civilization that, in a few thousand years of technological advance and a few decades of hyper-industrial and digital advances, seems to have triggered a now accelerating auto-extinction event). I will call this later, as if in an *arche-cynecist* mood and with de Man wryly in mind, the *literary structure of "climate change."*[25]

Now, to return to the opening dilemma: the problem arises not with "tropes" such as metaphor or metonymy but with something that is not a trope but nonetheless is "structured like a trope." De Man is in fact drawn not to figurative language but to what pre-emptively disfigures, perforates, deflates, compels evasion, or seems "structured like" but is not that at all. This thing-like status cannot be tagged since it is not a substitution of something else. It operates as anti-matter or a black hole might within tropes. One encounters this in those sentences in which a *negative enumeration* dismisses all of the substitute interpretive reflexes that the psyche or convention, philosophic ordinance or hermeneutic demands cast up:

> Anthropomorphism *freezes the infinite chain of tropological transformations* and propositions into one single assertion or essence which, as such, *excludes all others*. It is no longer a proposition but a *proper name*... Far from being the same, tropes such as metaphor (or metonymy) and anthropomorphisms are mutually exclusive. The apparent enumeration is in fact *a foreclosure*.... Truth is now defined by two incompatible assertions: either truth is a set of propositions or truth is a proper name.... [A]lthough a trope is *in no way the same*

> as an anthropomorphism, it is nevertheless the case that an anthropomorphism is *structured like a trope* (De Man 1984, 241; my italics).

Strange non-figure, that encircles the others, which is to say figuration itself (the entire software of face)—substitutive chains of all form of rhetorical elaborations we are familiar with, or leave unobserved, or are as if produced by: what, at their point of initiation as *anthropos,* or at least their being claimed (heliotropisms, metaphoric identification, mimetic passivities, the *eidos,* "light," the imaginary teloi of, well, eating more), freeze these in place, as in a vacuum, excluding "all others." Keep in mind, for Aristotle this means no women, no slaves, no children, and above all no "barbarians," those with other tongues—one must be in, and of, the *polis.* Moreover, it is, at its initiation, a foreclosure, no trope at all, since it has no anterior definition to cite or vary, though nonetheless "structured like a trope," a wolf in sheepskin, in no way possible to define as, in any way, "human." De Man continues:

> it is easy enough to cross the barrier that leads from trope to name but impossible, *once this barrier has been crossed,* to return from it to the starting point in 'truth.' Truth is a trope; a trope generates a norm or value; *this value (or ideology) is no longer true.* It is true that tropes are the producers of ideologies that are no longer true. Hence the 'army' metaphor. Truth, says Nietzsche, is a mobile *army* of tropes.... to call them an army is however to imply that their effect and their effectiveness is not a matter of judgment but of power. What characterizes a good army... is that its success has little to do with immanent justice.... (my italics).

It is an odd term, *trope*—particularly when at the point of its arrival it sediments, say, into representing what de Man calls "ideology." At the point at which linguistic consciousness (in this model a redundancy) emerges, it models hallucination's inability to tell itself apart from wakeful consciousness: in fact, the latter relies the more on iterative memes, citational *salutes,* tropological hashtags and place-holders. At this point the chain-gang of tropes morphs into an army, enforcers of what is untrue regardless. Yet this *other* of trope, "anthropomorphism," at once assumes the structure of and passes for a trope at first glance in the "enumeration" that is, instead, a *foreclosure*; "it" simulates the common form yet freezes it out, like an anti-body, refusing its life-blood, transport by substitution and identification. Again: the "proper name" simulates a trope without being one (in fact, having the power to *freeze* that)—like an alien egg snuck into another bird type's nest, it is sheer theft. It cannot be a metaphor for anything, since nothing precedes it as an iteration. This *technogenesis* and self-differing violence of language conventions (tropes troping) remarks the perpetually effaced installation, maintenance, and mutation of the "material" traces—Benjamin's "pure language," say—through which an artefaction of values, hierarchy, and co-optation is imposed (when, say, Constantine strategically adopts "Christianity" as a war god at the Milvian Bridge in order to deliteralize and delocalize a fragmenting yet still proliferating "Rome"). Inscriptions, accordingly, are like an impossible itch in the back of one's skull, where the projector projects, before which there is no artifacted memory or perceptual program—and if one were stuck in a "bad" movie, totalized or foreclosed within tropological systems?[26]

The irreversibility of a "proper name" that is blindly stamped as an exclusionary act and does not have to define itself—to whom, with that army?—echoes too strictly de Man's projection of a reading that does not or cannot *return* or *relapse* in the reflex of "recuperative" hermeneutics, of all varieties in which identification is recongealed, and which here serves both the imposition of this proper name (in this case, like the "Anthropocene") and its refusal to identify itself otherwise. This is peculiar, since the *anything but a trope* status of an "anthropomorphism" both appears as the *antidote* to tropological trances and, at the same time, something beyond their worst forms of *totalization*. The latter can be taken to accompany not just totalitarian memory controls (nationalisms, tribalisms, and so on) but the saturation by figural systems that have become false consciousness or false substitutes for what was not there to begin with ("ideology"). Elsewhere this appears as the divorce between *descriptive* language and the pre-phenomenality of any inscription—wherein "descriptive" language is not direct, mimetic, or indexing but comes into being, itself, as a defense, deferral, and concealment.[27]

If an "Anthropocene era" depended throughout its reign on something called anthropomorphism, the latter would be key to study or at least define. One wonders, again, why volumes were not devoted to this, the scene of a crime of sorts? De Man portrays *it* in a thing-like mode, at first hiding out among a herd of tropes, just one among others, yet contaminating and *freezing* them, in effect totalizing them into an army of power for what is, by then, a self-feeding vortex. It is very definitely not the same as personification or *prosopopeia*, and it has nothing to do with face (the anthropo-technic capture of trope). It is in one sense pre-emptive and lethal, a power that excludes and appropriates, arriving as a *foreclosure*. It is not a trope, nor a word, but sneaks into the parade of figures in Nietzsche's enumeration: it assumes an anonymous disguise as one among other tropes while remaining in a predatory mode. What is odd is that it seems reversibly doubled: there is as if a "good" mode (what is before or outside mere trope) and a "bad" mode (what freezes and forecloses). It simulates *trope*, which anthropomorphism is "structured like," but inversely has the qualities of an inscription—that is, what *descriptive* language and tropes are designed to conceal or efface. Its assertion is exclusionary and totalizing, to cover the fact that the proper name has no referent, no original to cite or copy. How could it be otherwise, when the effects of "consciousness" when asked to account for themselves do not depend on or derive from the animal or body that is or conceives itself as living? It cannot be a trope because it derives from nothing and cites nothing: it steals its place in line by simulation.

The claim is that *anthropomorphism* is non-existent, despite unnaming a mode of comportment that allies "the eye" with exponential *extractivism*. It does not project or express or metaphorize the world as narcissist reflections of itself and its own familiar qualities. It doesn't have any. It conceals itself among tropes, shares their "structure," but then freezes and forecloses a system in place, which, technically, cannot escape substitutive chains that appear simultaneously totalized and disconnected. The "anthropocene" is similarly a Trojan Horse figure, one that betrays the promise of an ex-anthropic perspective, the "promise" on which Western discourse has hermeneutically gambled with the equivalent of exotic financial instruments. The era of face in this sense was an artefaction and parenthesis, like that of a little ice age or a Holocene. It is an act of theft as a structure, hence the doubling between a "proper name" that conceals its lack of referent by violence and an inscription that puts the trance of figural language in hiatus.

The term "Anthropocene" projects itself as a future *inscription* that can only properly be recognized or read from a position after its collective disappearance—a logic that inserts a poison which freezes, totalizes, excludes. The *anthropos* or "human" (the Greek or Western installation) is never generalizable or universal or even applicable to something like a species or animal type of evolutionary phantom. It is a bandwidth, encoded. It is a group construct or "we": it *never* named all the hominid variants or even a living organism. Thus: "it is easy enough to cross the barrier that leads from trope to name but impossible, once this barrier has been crossed, to return from it to the starting point in 'truth.'" It is interesting to overlay this with the parallel movement that preoccupied de Man's last published lectures—an *irreversible* passage that, despite its grammar, would parry the "relapse" by which face, recuperative historicisms, and the management of anteriority (inscriptions) would be maintained. The "relapse" is the technic by which the dominion of anthropomorphism would be maintained, and ecocide acceleration guaranteed.

This is where defacement seems less to erupt than pre-emptively hum—but then it would not "deface" an extant object or face—as the totalizing turn against a "totalization." Not as some violence applied to face as one might troll on the web today to do a take-down from the shadows, nor as a resentful attempt to usurp it, but as a norm from which specific histories of face would be one commodifiable trajectory or story arc, one confused and consolidated by anthropomorphism—where that occurs or is produced and managed (television screens, talking heads).

One can transpose all this into an era of climate change—which is the time when there remains (as if *officially*) no climate change and climate change at the same time—when its invisibility is (in measure) structural and decreed. *It would imply, as de Man had figured out, but also as telemarketing algorithms have long since simulated and contrived to capture (and install), that resistance take the form of an equally totalizing defacement: the entirety of this foreclosure is delegitimized.* The poison pill of the proper name that has no relation to any originary *truth* and in turn transforms tropes into its subjugated and subjugating army, has effectively incorporated a dead-end technic, a "proper name." For an army of tropes to enter the field of battle as frozen or foreclosed, in advance, is to initiate this history with a zombie army that would nonetheless be taken, across this epoch, as the structure of time and experience. It is one reason today that, when today's "corporation" assumes its legal role as not only a person, but also a suffering subjectivity (the "Hobby Lobby" ruling), it goes unresisted aside from grumblings. It is when *faced* with the ecocide, which said army of tropes advances, that it is fully delegitimized and seemingly empowered. This is what is implied if an "epistemological critique of tropes" is inescapable. It is the difference between the position of the torch at the back of the cave influencing the movement of animal forms on the wall for the spellbound "audience" sitting before it—who identify with faces or animation, perpetuating a mimetic template or accord.

As it turns out, the cancer smuggled into figurative systems at play in grotesque substitutive traffic, then freezing and totalizing them, and then subordinating them, frozen tropes, to its blank militarization—*is* anthropomorphism. It does not, moreover, represent in itself some species entity at all, the fact that it is welded to the hominid speaker or claimant is as secondary as the "personhood" now legally and magniloquently bestowed on some animals and corporations. Whatever is projected by this proper name it could not be called a hominid, or humanity (a trope), or the essence of human being, and so forth. There is nothing in advance specifically "human" about it, again, were that identifiable. This is confirmed by all the warring tribalisms today, the regressions to mythological histories as cover—one or another "we'" channels medieval Russia or, inversely, Mao *lite*, or for the U.S., some "age of affluence" is evoked (since it cannot recognize itself in any past neo-feudalism and medieval kleptocracies). Moreover, one reaches back for twentieth-century memes and slogans as cover (if not thirteenth-century ones), an acid reflex or robo-relapse shimmeringly anaesthetized before twenty-first century horizons under the new pressures of visible and invisible climate wars—since "resource wars" doesn't take account of collapsed agriculture, *megadrought*, mass climate refugees, and so on. When Nietzsche launches his strikes against being "human, all too human," and proceeds to write a rhetorical machine thinking itself outside of that, he does not mean some biological being or hominid—he does not want to erase the organism. It is about altering the proper name, and definition, of what mobilizes and inhabits that product of the first (perhaps last) several millennia of writing and hyper-technics.[28]

The import, of course, is that the *proper name* ("anthropos") has no definitional relation to any idea or model of "the human"—even though the latter could not be posited outside of something like it. There is, here, no first or previous term to cite, repeat, verify, or express a "human" essence since the assertion of the proper name freezes tropes, negates and militarizes them (there is no thing to be a metaphor "of" here). That means that not only does *anthropomorphism*, if it occurs or exists, not mean one thing but that it asserts by way of exclusion, guaranteeing there is no such thing as the "human" as universal humanity, as other than competition or conflict—as we see geopolitics reverting to the norm of today, returning to a pre-Enlightenment and neo-feudal template of hybrid and multi-planed "borders," contractor armies corporate and stateless. That is, what is called "anthropos" as proper name, now projected as a future anterior inscription, is anything but a natural or even species reference. The word "human" is not descriptive nor a word itself, then, but is always a verbal construct created as an exclusionary foreclosure. One can see, in the era of climate war, that this term will break into innumerable micro-tribal, territorial and corporate enclaves—including towns or counties competing, again, for water. [29]

Thus the proper name "human" never referenced hominids or even a living "species." It is a weaponized construct predicated on exclusion, one whose import is linguistic, inscriptive, artefactual, much as the proposed moniker *anthropocene* itself. It speculates on a non-existent "we," banked on, bound by credit and hierarchies. Worse, the era of climate comedy coincides with the withdrawal of the post cold-war geopolitical accord. The very thing that would universalize the prospect of a joint species perspective, "humanity" as the "humanualist" hominid that anthropomorphizes, guarantees the opposite: its fratricidal fragmentation—since anthropomorphism can only operate as an exclusionary usurpation. And it may reference what we take consciousness to be, without entirely reflecting whether "consciousness," if definable, were necessarily hominid or might be generated and exceeded by A.I. networks (to supposedly assure against anthropocide, future robot marketers speak of guaranteed blocks against "consciousness"). What if this future A.I., rather than the hyperbolic manager of data streams and exponential calculations, took an interest in reading?[30] It would immediately be *very* close reading and indeed unflummoxed by robo-hermeneutic narcissisms. It would not anthropomorphize, obviously. It would not be perplexed by metaphors only we "humans" know, usages that cannot be replicated, voice and affect inflection, cultural citations—and so on. It might get all that and move straight to the "material" play of forms, insignia, interpretive canons and epochal appropriations and their constellatory relations to the hacking of the real.

4. Brunch in the Ideovomitorium—or, WTF: Was "Anthropos" ever in fact, er… Greek?

Here is the painful part. Of course, there is one other possibility. How to say this—it's almost embarrassing. Well, let's appeal to *anthropos'* vanity. Since proposing the proper name Anthropocene for an epoch that above all signals its forgetting, the geologists unreflectively signed on to this land-claim and proprietization of the West's arc and nomenclature, capitalizing on its Greek splendor. It rests on the genealogy, embellishes itself, stakes a claim. The one thing *anthropos* swells at with a bit of pride is, nonetheless, his techno-genesis. His DNA for the British Geological Society is—you are, hyper-modernity is—after all, Greek, no? It is (in) his *name*. But what if… there are Trojan Horses within and for "Trojan Horses." What if what we call "the Greeks" were, so to speak, *trojaned* in advance?

A confession. I am not one for twilights—too romantic, murky, too nineteenth century, a time of whale-oil-lamps when they still referenced diurnal cycles. "Nature." What about daybreak? Who is to say it is not just a sheathing black "light" now indefinite fore and back—like the normal day in Shanghai. That is, before *anthropos* was installed, set in motion. (Check with your inner *anthropos* still on this—the one that changes the subject, murmuring no, calling *X* cynical perhaps, gathering its mafia about it, pointing again to its preformatted "we" to come, and mistaking all that for *esprit*.)

Just a little note, returning to the very serious mode of climate comedy whose operatics and acrobatics could busy one for decades, and will, since the new normals that unfold daily allow the carpet to curl up after each step—occluding the rupture or disconnect. Is it now normal to expect mega-drought, mass extinctions, video beheadings, hybrid wars, fixed "markets," guaranteed ecocide—yes, sure, perhaps, whatever. It will not be long before a next generation enters taking all this for granted, as the norm given, with no recall of an imaginary before. The resistance to climate change, like that to "theory," which is to say that to the *hypo-script* that traverses hyper-industrial algorithms, old textual accretions, and the encoding of cells and life-forms, is hard wired to robo-proprietizations. The ecocidal accelerations would have less to do with CO2 and emissions, global warming and biomutation, than with the relatively lame cognitive settings, installed, exploitable for honing techno-power, fossilized and lingered overlong in a sort of accelerated stall. But to say "anthropos" never existed as an organism is also to say, or by way of saying, that he never was installed as such either. This calls for a back glance, a bit of detective work, a good joke on us. Back to the Greeks bearing the gift of "anthropos"—scene of an obscure inscriptive crime or "swindle," which indicates the corporatocracene was active in the forging of an archival template.

There is a famous painting by Raphael, "The School of Athens," where the event of this gift is economized—all the boys are gathered in some signature gesture, perfect for appropriation as a cartoon inscription. (Of course, one would have to say, cartoon characters and pop cultural seizure begins, and to some extent ends, with Plato's "Socrates.") Raphael gathers the philosophic mafia that launched the flowering of Western thought, mathematics, and descriptive science and would be refolded, reclaimed, redistributed, expropriated by its declared re-birth (*Re-naissance*) and re-initiation. It is comprehensive: there is Parmenides, there is Euclid, and so on. And there are Plato and Aristotle at the center in full stride, with *finger* and hand pointing—finger up (Plato), palm down (Aristotle). (We will omit focus on Plato's digitating and deictic finger, already pointing, invading, protruding, staining the mighty strides, up, down.) But then, there is another figure included, lying on the steps, the only one horizontal and pointedly ignoring all of the rest, and he is just reading a book:

Diogenes. Raphael seems to mark this counter-point, leaving him central yet altogether apart. No marching, no attention granted to the militarized group. The roaming icon of "Diogenes" is unaccounted for, well-smudged in late modernities with the curiously foreclosed metaphor of "cynic." He is elsewhere pictured with his dog and his lamp, carried in the daylight, looking for a real or honest *anthropos,* that is: looking still amidst this troop of titans for "anthropos" at all. It gives pause and opens an alternative history within those layered about this painting's über-canonization. It would take someone more cynical than myself, if that is the term, to really suggest, say, that "we" have just undergone a financially engineered species split by the same folk that streamed climate denialism to the masses while knowing the opposite, say, or that even discussion of "inequality" today is a distraction from this. I could never say such things myself, or rather, my inner *anthropos* might thrash about at first and need, at least, his meds. But I return to the question posed, this holiest of premises: is the Western *anthropos* in fact even Greek, does that anoint his legitimacy, give him a proper imprimatur with which to leave, even, the name *Anthropocene* stamped into (or after) an ecocidal Earth that he claims disappearing mastery of compulsively?

It is in Nietzsche's *divertissement* or "recreation," *Twilight of the Idols,* that we get, again, the gesture that puts the Greeks as received (as always) into question, and it is not the sort of attack on Plato's crafting of "idealism" (finger up) that would be overturned or whose overturning Heidegger would artifice and seize on to keep metaphysics in play in order to slay it again and again. It was never *in place*:

> And please do not bring up Plato as a counter-example. I am a total sceptic when it comes to Plato and I have never been able to join in the conventional scholarly admiration of the *artist* Plato. In the end, I have the most refined ancient arbiters oftaste on my side. It seems to me that Plato mixes up all the forms of style, which makes him a *first-rate* decadent of style … The fact that the Platonic dialogue, this horribly smug, childlike type of dialectic, could strike anyone as charming—this could only happen to people who have never read any good French writers,—like Fontenelle, for instance. Plato is boring. In the end, I have a deep distrust of Plato: I find him

> so much at odds with the basic Hellenic instincts, so moralistic, so proleptically Christian—he already has 'good' as the highest concept—, that I would just as soon refer to the whole Plato phenomenon in harsh terms like 'higher hoax' or, if you would prefer, 'idealism', than in any other way (Nietzsche 2005, 225).

Plato was already a "decadent" *literateur*, "boring" stylistically, crimped by Socratic play-school ethicism—at least, from a certain perspective among many. It simulates Diogenes' view at the time, his life spanning the generations in Raphael's set piece. Living out of his broken jar, carrying his oil lamp in the mid-day sun, Diogenes wasn't buying the installation of the Western "anthropos." He harassed Plato as a soft romanticist whose version of Socrates was a joke, disregarded Aristotle as a codifier, and famously dissed the latter's conquering student Alexander the Great as he literalized the pop program (implemented it) as world transforming *empire*—allegedly telling him, when he is offered any gift from the new transformer of worlds, to move aside as he was blocking the sun. Eclipse, dis-eclipse. When Plato offered a working definition of *anthropos* as a "featherless biped," Diogenes rushed in with a plucked chicken (*ecce homo*). And yet, the familiar of *these* men, and others of the "school of Athens," Diogenes went looking for *anthropos* famously, a real or honest *anthropos*, without any luck. Moreover, he did it with an *oil* lamp held to the daylight sun, cancelled by the latter as "light," but also indicating the latter to be a technic producing "light" itself (not, as in Plato's candy for plebeian readers, "the Good," the "idea of ideas," the Father, the whole heliotropic armature). Of course, there is the oil in the lamp itself, even if from olives, anticipating the petrol-cene catastrophe of exhumed and hyperbolically consumed "stored sunlight" (fossil fuels, oil—remember weaponized "Greek fire" that was a trade secret!). And that lamp cancels in advance the Platonic "cave's" management and casting of a cinematic model of "consciousness"—flickering against the wall, for which the true light would be, we project or double, outside. Diogenes (the residue of the term "cynic" does not get us there) practiced a defacement of the entire project of installing "anthropos" at its inception, as if he could fast forward in a cinematic forecast of the coming several thousand years arc of this installation and its ecocidal program. If, today, a certain *we* can only give a back-glance at an "anthropocene" era by virtue of its having been given a proper name, it would seem Diogenes did the same at *its* purported confabulation—called a "Socrates gone mad." But more to the point than Diogenes' practice of epochal defacement in advance, his complete pulling away from the artificed "we" then being consolidated is mimed in the inaugural act of counterfeiting, of rendering the currency fraudulent, that earned him exile status and inaugurated a different cosmopolitanism (of which he is said to be the first). At issue in his *archecynematic* trolling, or his dismissal of the "we," is that he still could not find "anthropos" across these generations that were installing its premise and promise. Raphael's propaganda pop art brings the Greek gang under Christographic policing, thereby inventing it.

Thus, the last possibility mentioned above: that "anthropos" was never installed at all, apart from his mimic adaptability; that he not only never *existed* (and obviously didn't name hominid variants in general), but that "he" is or was not specifically and irreducibly "Greek." Beware, indeed, *gifts* bearing "Greeks." He would arrive not with cities but something else, a *telepolis* composed of his ilk—linking the definition of politics to his arc, detour, incessant eating, devastations. This last prospect would imply *not* that the material events of these works, lives, writings, archival coups, and imperium (the whole known world, for Alexander) were the technogenesis of what we find, today, devolved to Last Man ecocidal hyperindustrialization "culture" stemming from the Western adventure and its American cowboy distillations. It would suggest that we, that this "we," is perpetually invented and consolidated *through* its own production, at every epistemo-aesthetic juncture, in every redaction through this family line or algorithmic setting of a perceptual "present." It would be droll to suggest, in crit-speak, that this is done by a sort of hermeneutic reflex and normatized "relapse" all but pre-programmed in, added in to freeze and totalize all but on its own, *Autos*-matically, automated as a sort of acid reflex of cognitive mechanics and referential encasement. It would process "the Greeks" as canonical misreadings ("Platonism" for Plato?) that confirm and induce the trance-like hashtags for reference and neural response, a perpetual smoothing out today policed by bots and algorithms, "ideas," tele-adverts and pre-emptive data-harvesting, mimetic management of the screen, and the hollowed out digital theologies. *It* would be independent of its host organism. If so, then we cannot even point to "anthropos" as our legacy and pride, our twilight and our narrative base. It would be something else, for which the Greek legacy served as a front rather than index. What would that be?

The proper name "Anthropocene" arrives after and as its own twilight, which negates or occludes the fact. We donate the natural image of a cyclical twilight—the pathos of going under and the promise of midnight passing—to what is a flickering screen, where an epoch of the *cin*-anthropocene terminates and the trope of an "eye" passes into direct neural implants, captured mimetologies replayed for purposes of guiding consumption and anaesthetizing the nostalgia for an imaginary *polis* at all. "We" are left, so to speak, with the residual prospect or active remainder. That is, that not only did *anthropos* never exist as a specific organism rather than a linguistic-mnemonic regime, but that something else had been at work which we don't have a proper name for, or any direct or phenomenal evidence of. Rather, it would not simply repeat the three-card monte of *anthropomorphism's* claim as de Man accords it for being outside the tropologies it passes for and pretends to number among—a claim of imposition, abrupt force, exclusion, of freezing and foreclosure, enforced by an "army of tropes" or, we might add today, bot surveillance. *Rather, "it" operates like a reverse anthropomorphism.* It goes into reverse. It is, then, misleading to call *anthropos* a "failed prosopopeia," as I feign to above. Per definition, at least in de Man's account, anthropomorphism precedes and forecloses figural systems including *prosopopeia*; freezes and then manages figuration itself *as if* from without. It would be impossible to personify, animate, give voice or face to what precedes the latter's possibility.

Perhaps we chose the wrong drugs from Plato's pharmacy—the ones we gulped (the verb *eidein*, for instance, which fused knowing with seeing) were meant as trial packets to discard. We mistook cartel street brands for Platonic Big Pharma. It's one thing for Nietzsche to trope an overturning of a Plato who never was, and another for Heidegger to commodify the gesture passive aggressively, to divest Nietzsche and own it, in turn, when it was never there to begin with. *Metaphysics* would have been something "we" create, re-iterate, consolidate, generalize, and paste into place ceaselessly and addictogenically, that feeds the "we" in concert with accelerating ecocide. The latter is called, say, anthropo-narcissism, but it is presumptuous if, as in de Man's account, the first and key term looking at the mirror had no presence or face (as is said of the amnesiac cipher, Gregory Peck, in Hitchcock's *Spellbound*).[31] This might return us to something like a *literary* structure or, perhaps, cognitive program of ecocide that would have no necessary genetic link to an *anthropos* which or who never arrived quite to begin with. Something without a name, neither "human" nor of *anthropos*, without a model, itself the inversion of any *anthropomorphism*. It might seem that "we" got stuck in an aesthetic regime (did somebody say, "aesthetic ideo*graphy*"?) that at some random juncture proved more economical to consolidate and weaponize as a generalized regime of reference, perceptual calibration, visibility, invisibility, animation—and stuck.

Let's trace the fallout of this. If *anthropos* were not installed with the Athenian archive, that is not to say it is or was, then, Egyptian, or even Hebrew—sorry, Joyce, no GreekJew neither, though whiffs of a psychotic theology remain. It is also not to say, as Nietzsche does, that we get "it" really from Rome, a missing "birth" consolidated by its re-iteration or, as with Raphael, a *naissance* posited only by its *renaissance*.[32] This prospect could not even be called a defacement of *anthropos* the non-existent, though it does strip a few epaulets away and retires the genealogical pretense with all its baubles. Ostensibly, this would parallel the urban myth that "life" was not terrestrial but carried in by some contaminated asteroid or debris from, say, Mars, that "we" are the aliens—or that the grammatical "we" is itself.

De Man's last intervention is very difficult to "relapse" from. It is unfair to call Anthropos, as I pretend in the title, a failed *prosopopeia*—an uncalled for slight, no doubt, particularly if all figuration would technically follow, be frozen and foreclosed, or contained by something like anthropomorphism. (It could be called a failed *anthropomorphism*, but that seems a tautology.) The sentient apparatus, referential codifier, and mnemo-technic regime that was sometimes called language in the twentieth century lens is what we associate with his claim to utter distinction (let's still give "him" his gender imprint for all the obvious reasons, mindful, nonetheless, that the apparatus itself is not obviously gendered). This sentient being may immerse or wrap itself in figurative language, let itself be creatively shaped by local "cultural" imprints, but only to the degree "he" forgets that this latter occurs always in effaced quote marks. Figuration is sustained in full play and substitutive dolphin flips, inaugurated *as* it is frozen, first, by the faceless coup of *anthropomorphism without anthropos* much as it is, today, by the digital and techno-controls of mediacratic and dark-net streaming and data pillaging patrolled by bots. Thus today the errant revenants of Hollywood: zombie inundations, unfriendly machines, para-humans of all varieties, pathetic "superheroes" in knickers, demon teens. Whenever what we call consciousness dissociates—I am actually *here*? this is really occurring to *me*?—or whenever one reflects that the world before one can be derealized, as if another's or a cinematic screen, the rivets pop and the forged contract flashes up. Every time a financial commentator remarks the impossibility of a pretend and extend game in the megadebt economy of extinction, then returns to the casino; every time *irony* gets named as the perspective of the non-"human" itself; every time climate change is derealized as a burdensome invisibility nonetheless before our eyes, this disjuncture flashes up.[33] But if *anthropos* had, all along, a certain affinity with the headless and bodiless corporate personhood, one can see why "Asia" or "the East" pulls away from him, loudly and geo-politically, as a sort of contamination: *he* no longer even seems like *anthropos* of old, which the "others" were coerced to recognize, love, want for a while to be. The "2008 financial crisis"—wherein *he* was shown to have no connection at all to his own protocols, not to mention the Bush wars, or the ecocidal acceleration of earth—delegitimized these fictions. But if this never quite was *anthropos*, what exactly defines *last man* culture as it presents itself?

One can compare this, parochially, to the difference de Man asserted in contrast to Derridean "deconstruction": there never was any "metaphysics" to begin with to deconstruct. Anything akin to a text deconstructs itself in advance. What we retro-project or narrate as a "metaphysics" as if to *overcome* would be ceaselessly produced in and by a perpetual non-present of narrative redaction, of editing the "rushes" of the day—which would return a certain responsibility and choice to a present in which human civilization technically *could* have arrested a cascade of mass extinction events, except, for the same reason, not. Ecocide would have been inscribed in the technic phantom of the home (*Oikos*), the interior or reserve, proprietization, the perpetual exclusion of "all others," the grammatical "we." There would have been no metaphyics to deconstruct, outside of the perpetual relapse by which a "present" were secured, artificed, foreclosed in its redaction; text does all that in advance, and it still ended up, for us, masking its status as event, accruing, deferring its readability, totalizing its tropological *bourses*, "capitalizing" things, creatures, resources, bio-data, until it crystallizes—at the end of the era of cinema—in last man ecocidal totalization. In the pan-metaphysics of the digital foreclosure itself, past tipping points, of the "last" anthropos. De Man is quite explicit from the beginning of his thought experiment: it has been thus since there was not this or that author or oeuvre, but since "text" of any sort. Before the Greeks, before and including what we pretend to name as "Capitalism," as if that were out there, and the clever antagonist were purer than that (such is the immersive trope), against *it*, if perhaps unwilling to trade in one's iPhone and retirement benefits to make the point or prevent any particular genocide. Before "Capitalism" would be an *extractivist drive* linked to the settings for sight, "the eye," the hunt, the purity of a techno-theft associated with fire, echoed on Wall St. today, as if the end of "the cycle," returned to its nominal site of initiation, technics as unrelenting and mimicking Promethean theft. The first megafauna to appear on the cave walls would be the first to become extinct, at the scene or advent (already highly advanced in the cave walls that remain, for us, to read—likely late, decadent or modernist ensembles themselves).

At a time when the so-called "humanities" is being globally defunded as irrelevant, and reading closed by data-streams (or foreclosed), the question resonates of what it had been doing—and of course, why its stalwarts missed its opportunities, predictably, relapsing into a trapped vortex of historicist, "political," humanist and more or less recuperative tropes. Of course, that even the Pope can muscle up today, while the farmyard of disappearing tenured professors in the humanities, the nominal "intelligentsia," cognitively resist taking it in, still partition off, occlude it, nor publish a joint manifesto on the subject, is endemic—which would have included, indirectly or by design, the late Derrida.

This returns to what I named at the opening, deadpan, a literary structure of climate change—that is, the playing out of ecocide by way of cognitive settings and memory regimes which surround and advance extractivist ecocide into the hyper- and post-capitalism, klepto-anarchic gaming and global corruption of "we's" and mafias today. How did a hermeneutic regime that preceded and mimed both the capture of aesthetic production (aesthetic ideography) install itself in the acquisitive and taxidermic "relapse," reflexively and mechanically ensuring this trajectory. To some degree, the *elan* of trends today like *Object Oriented Ontology* reach for an animism impossible to locate, after all projected or accommodated or ventriloquized "others" have been run through. From the perspective of this taxonomic and romanticizing reflex, an addictogenic "we" is projected and weaponized, visibility honed to target, the artifice of the "senses" streamed or calibrated, "life" re-engineered. The house of anthropos' self-inventions along the way—say, Freud's giving them an "unconscious"—serve more as outlets than modes of self-awareness, naming and containing spillages. The mergence of the protocols of the "Enlightenment" with assured ecocide mime Freud's tracing of a "death drive" that is no drive at all but the logical exteriorization of a parenthesis whose cover and propaganda was the home, the enclosure, the *Oikos*, the Self, the interior reserve, the genealogical fable, the recuperative interpretation which traverses both hermeneutics and the everyday technogenesis of what we call, figuratively speaking, "consciousness" and "perception."

So, then: *Anthropos*—master of "life," destroyer of worlds—is not even Greek (then again, was that not implied by *The Birth of Tragedy*?). Let that settle in: you will not have the residual dignity of having been auto-erased by and within the noble trajectory of intellectual and techno-progress that the West's imaginary presents. Even Žižek, as Derrida did, rushes back to the "Enlightenment" parapets before the spectacle of barbarian approach, the trolling of ISIS, as the two nihilisms face off. Then what? Was it set in motion, as Timothy Morton opines, in the opening grid-work of *agrilogistics* (Morton, 2015), or the initiation and percept coalescence of cave "paintings," or rather, *arche*-cinematics before any letteral or pictographic script—yet where a regime (among others) of light and shadow, of movement, of the "eye," of mimetic animemes, would coalesce (someone, some priest, would seize the torch), the collective then repeating these *arche*-hunts as identificatory models in the quest for food, energy, protein, trinkets, megafauna.[34] Somewhere in these early processes, or many times over, in a "prehistoriality" that is, nonetheless, hyper-industrial still, a certain inertia of interpretive organs and perceptual bandwidths found itself useful to political forces to lock in place or hone: reference would mean their communal repetition, mimetic tags would substitute for abyssal processes, gods would shift from animate projections of unknown forces barely personifiable into narratives or statues, fixed, or rituals of hashtags and emojis. This, you can say, is where the "still" is invented—not before cinema but after and to fix, as temporal marker, an *arche*-cinematic flux that precedes all script or even "image." This, you can say, is where the apparatus that works like a sugar hit on the consciousness-effect it rewards (in projecting), parasites the group, the organs and organisms of the cave, the referential stabilities predictably useful and data-fied. It is here, one could say, and hence where "text" occurs, that the relapse is installed, the backfold, in deference to a "we" frothed up by the priest with the torch, and the mnemo-technic transposition of inscriptions into the info-bytes of assumed presents and the claimed immediacy of "my perception" cited, formalized. And it is the inertia of the technic acceleration that takes over and protects this artificed enclave—the "magic tent" of stacked branches that Justine (Kirsten Dunst) and the futureless boy of Lars Von Trier's *Melancholia* retreat into. In short, the very template out of which *anthropos* would take and presume the real that he knew how to handle and name, dissect and shadow-graph, and which rewarded "him" ceaselessly with the sugar-hit of yet more proprietization to come, and the phantom of "consciousness," would take off on its own, advance by turning upon its own traces and cannibalizing them, and present itself, in the knowledge base and in the daily, hourly, momentary redactions of memory—allowing a self to be posited as a *still*, then projected, then quoted, all as if instantaneously since the gaps, intervals, whirring gears, and additives would be erased and occluded as part of the bargain, this apparatus could and would be applicable to any textual formation or event, whether called canonical interpretation, history, or hyper-industrial victimage. The last man would have been with us from the "first," balanced by the ardors of war, hunt, pillaging, erasure, and weird gods (nowhere as weird, though, as the monotheist anti-God of alphabeticist grammar, the precursor CEO of the bodiless corporation). Here is where the "apparatus" of Hollywood is first installed or effaced, and it operates as inertia, as a feeding mechanism that counterfeits a present, a "still," for its organic host, defrays the inscriptions that precede projection and the "public space" of the cave wall or screen, gifts the hallucinations of empiricism, historicism, pragmatism, nationalism, literalist tropological systems, frozen vortices without exit from substitutive relays, the "real" hacked in advance of itself and subject, in the digital transcription, to calculation and capture. What "the Greeks" imbibed and quaffed, exempting Heraclitus, and what would be pop-cited already in John 1:1 as, simply, the "beginning" (bureaucratizing any *fiat "lux"*), *logos,* was already a Trojan Horse—the corporate theomorphism enabling pre-industrial framing devices. By what necessity, after all, would there be only a "*Four*-Fold"? Grand theft *Autos*—by itself.

These coalescences of an autonomous seeming *apparatus* (forgive the improvisatory term), a machine in the ghost that pretends it is the ghost in the machine, kept even (even?) Marx from thinking "materiality," and compelled him into a Christological narrative. And it is this which says, or ventriloquizes, not only the "I" of "I, Anthropos," having rendered the host being compliant, but the "I" that cites itself into Hollywood credence on a sentence by sentence basis as the (Western) House of the Anthropoid—in exchange, it gifts the latter denizens with the imaginary of an *Oikos.* It is entirely detached from the fortuitous archival catacombs and structures, cities and architectures it devises, including their hyperindustrial accelerations into full service digital streams and mnemonic implants to come—that is, the era disclosed after the exteriorization of cinema has run its course, and left consumers with superhero franchises.

Thus, *anthropos* would never have been Greek at all. And whatever says "I, Anthropos" is speaking not from the wealth of a human historial experience, and not from the citation of a legacy of civilized advance and mastery of "earth" (er—megadroughts, mass extinctions, exponential ecocide—mastery, really?), but is, in mute compact, allowing this obscure reflex and device, hidden within the intervals of mnemonic editing (and confirming the product alone). It might be called all but automatic—and, moreover, the exteriorization of that Greek word, *Autos,* out of which the panoply of grammatical positions, of seeming "subjects" and "we's" to initially proliferate, defend themselves, exclude all others, freeze the very tropological play they depended on for credit and a definition of "life," even to confirm (falsely) that that speaking "I" was, technically, alive itself.

It is hardly accidental that, today, after "cinema studies" had made such progressive use of the term *cut*, assuming that it taught and reframed our continuous consciousnesses with the lessons of interruption, that our specialists have decided that the brain works with this "cut" incessantly and initially—that the continuities are what the redaction papers over, that no cinematic product can mime the nano-mnemonic gum and drug that yield the product called "consciousness," that even the latter was not a "human" trait (alone or otherwise), but the implant of a semio-biological exfloriation and paleo-corporatism that would saturate and anaesthetize that organism or "brain," until the host was used up, the *Oikos* fully exteriorized as a back-projection, a fully policed *Nachkonstruction*, the disconnect with things and creatures and terrestrial perceptibility complete in the circuitry of a *telepolis* that cannibalizes all the dead organic life previously on earth (fossil-fuels, *oil*). Plato's cave indeed—at least, in so far as a proper name like "Plato" replicates the containment gesture. It would seem, then, or the notion need now be entertained, that *Anthropos* bears the Greek cloak to backstop and name an initialization, an HFT software given to end-running itself, in the name of power, that it had all along been an "it," and that the gift of the illusion of *prosopoepeia* to the hominid type was an agreed upon drug, poison, *pharmakon*, or *Gift* indeed, for which the addictogenic creature, so richly rewarded for the arrangement (smartphones, Nikes, treats of all kinds and girths of unprecedented proportions), could not possibly stop. It was an amusing error by our Enlightenment rhetoricians and liberal utopists, or for that matter scientists, to assume that the prospect of extinction events, convincing mathematics (the import of crashed foodchains, mega-drought, "population culling," resource depletion, and so on), would evoke a response of unified "enlightened self interest"—when the "self interest" in question would be that of this apparatus, for which "Capital" seems but a tool and tropological red herring. When, today, speculators identify the sociopathic or psychopathic traits required by the CEOs of predatory trans-nationals and megabanks, who were to keep the system running by creative destruction—and who, at the right moment, gamed even that in turn by channeling the "secret sauce" of hyperbolic mega-debt—they are perhaps missing where just these lords of the fabled ".001 %" are less masters than tools of an acceleration for which their hoarding would, necessarily, be temporary comfort, buy a few more decades or generations for the genetically engineered survivor caste imaginary.

It turns out, then, not that we need mourn the disappearance of *anthropos* or the extinction of life forms he, it, regards as tools and energy sources, since "he" never existed as a creature himself. *He* was neither specifically Greek, nor Hebrew, nor Byzantine, even if "he" were honed as a type within the geographematic enclave of the Mediterranean and the bio-semiotic competition of numerous tribal variants in a happily temperate earth-bowl during a lull in the climactic action of a spittle worth of millennia. Well done. But somewhere along the line, this apparatus gifted to "him" a certain longevity and imaginary and selectively real hegemony that came with a full tilt ecocidism baked into its technic inception—what appears around us fully today with prattle of singularities, the use of the body as a prosthetic to redesign or leave, the imaginary adequation of the computer to the "brain." (It is out of this blurred imaginary that one wonders, say, if the Neanderthals were the good guys and progressives—assigned the propaganda label of sub-humans?) One is only allowed to troll this "history" for the moment, with a twinkle in one's so-called eye, now that it is, in a sense, encircuited. That is, now that the artificed "Oikos" has fully exteriorized itself and reconsolidated as a tele-politic circuit, in which the remaining acceleration will now take care of itself regardless. One cannot ignore what tipping points *passed* means. Among other things, that all geopolitical and "breakaway civilization" acquisitions, all Hollywood trends, and all mafia decisions regarding resources and wars, indeed, all inducements of "the public" to remain in the same aesthetic ideographical trances that have been digitally mastered and replicated, occur, "willy nilly," under a twenty-first century politics of managed extinction going forward. Disposable nations and peoples, climate refugees without nation or earth to land on (the Rohinga are one of the first of this new breed), corporate mediacratic streams, the requirement to be as close up the pyramid as survival status allows, decade for decade. A few centuries is nothing in this calculus.

And this *is* "daybreak."

Perhaps these new horizons, irreversible, present another gift. Not in the politics of managed extinction that hums in the background, per definition. Rather, the time now opening promises without promising to be one in which the most interesting narratives in the human chronicles remain to be played out or written, *without promise*. It is entirely bogus for our robo-leftists to squeal that this, acceded ecocide, means nothing can or should be done or matters, and so on. In the twilight phase of anthropoid earth, the true hyperbolic adventures and aesthetic creations, the relinquishment of the proper name altogether, and the greatest pleasures on a crowded and burnt out "planet" (in our schemata and chemical or molecular compositions) proceed to play out for a sufficiently representable group for now. Daybreak—of sorts. And it cannot be incidental that something like an *arche-cynecism* or *cynematicism* attends this feast, again, circling back, and not only in the simulacra of the digital swarm. Moreover, in this polar vortex of imaginaries, it is that latter's disavowal of "the political" as an anthropoid contract that marks its identification with the disavowed of the world, those outside the anthropoid economy that is heir to the colonial epistemologies or perceptual regimes. All of this, one should stress, might seem "natural" enough, as there were no guarantees that techno-civilization did not have a sell-by date, that it had not been luxuriating on borrowed time since, say, the Cuban Missile Crisis, that a window or parenthesis would not close—without tragedy, without exceptionalism, without telos, without (Hollywood) escape. Contrary to the gurgle of discomfort from ethicist and political imaginaries and their "beliefs," neither the new ecocidal sublime nor its arche-cynematic lens conforms to the dismissive epithets of pessimism, apoliticism, and so on, the reflex of the cartel "we." It does not result in some exotic "nihilism" activists must ostracize to proceed, utopists write, political prognosticators churn on. It just removes the illusion that there is a telos or open time-span or comprehensive future, or that the "promise" represents any more the temporal premise of speech acts. Moreover, since numerically the abundance of human lives of the present swamps and will swamp the marquee populations of the so-called past, and the supported populations of industrial economies and black economies have never experienced such consumables, transport, pleasures, energy consumption—one might consider this period or time bubble, our own, as using up this reserve and that of future populations numerically, exponentially, and leaving the tab. Point: in fact, one might hold the greatest expectations for human types and experience, adventures and narratives, wars and erasures going forward. It should be extraordinary. It just won't last that long by previous metrics, like the cosmic mayfly opening the Nietzsche text taken up by de Man above. One of the street jokes of climate comedy, surely, and thus too embarrassing to trot out usually, is that all the forecasts and global rhetoric focused on evading "tipping points," all the calculations of what minimal success might even be once the numbers are pared, gamed or doctored, the definition of making it ends abruptly at the end of this century at best—then the screen goes entirely dark. Beyond that is not even imagined. Hilarious.

~

Here at the Ideovomitoreum there is time to pass waiting for brunch. It is a strange form of time. The morning haze tends not to part, rather like a Shanghai day—nor the odors. I will let you in on an amusement I occasionally pursue as we wait. It sharpens one's pre-Cog skills. We must learn, after all, to live without "anthropos," difficult as that is to imagine, and that includes the robo-*we* that worked so well in tribal survival and contexts of pillage or proprietization (theft), but now appears, well, let's just say its levers got stuck, or gummed into place, and the anachronistic settings and memory regimes—not technically "believed in" by their carriers—appear to rattle.

I'll share my own way of passing time. In a manner of speaking, I keep a pet "anthropos" tied beneath the sink for this ritual—a more or less pure breed (less fuss than, say, Collies). And I toss it treats, nuggets, conceptual softballs, horrors, aporias, from time to time to watch them get snapped up, turned into a sort of rationalized candy drip, lips smeared with an infantine or eerie satisfaction each time (it reminds me, for some reason, of a monitor lizard). Most of all, I like to see not the predictable rationalizations, the invariable proprietization back into some inaccessible *me* or *we*, the hermeneutic three-card monte, followed necessarily by some weaponized belch of promise, hope, selfie-ness, victimage, revenge, and so on—usually good for a cognitive smile for the student of failed prosopopeia (that is, "prosopopeia"). Call me decadent, but I find the faux aura of satiation that attends the sugar-hit of artificed cognitive relapse touching. I keep it chained not for fear of its natural aggression (one would rather say, passive aggression), nor of contaminating bacteria from his odd digestive processes. He takes it as empowerment. You can think of him, if you like, as Odradek's loquacious cousin, the family embarrassment—yet he has a secret. In between feeding and treats, you can listen to his murmurs and mnemonic reflexes: the *we* to come... the "singularity" to come... the *deus ex machina* of "technology" to come... the "ethical"... "absolute hospitality"... (How the Left got pulled into this, unwilling to sacrifice any one or anything, obverse to their twentieth-century forebears, is anyone's guess—my own is Marx's cynical inversion not of "Hegel," but Christian teleology, a rhetorical seduction for "the masses.") But of course, he is not the "anthropos" of old. He ceased thinking death, nor does its tissue matter, chemical make-up, body, energy hook-ups, attention span, or robo-defenses have much in common with *anthropoi* of even a half century ago and much less. Moreover, if you shine a light on it, if it goes silent for too long or starts humming to itself, you see something startling: its ribs still show—in fact, the body cavity is pretty much hollowed through, all those items of ingestion fallen to the side, often smeared but intact. Even so, he belches.

I sometimes forget about him. He's always the same when you return—emanating confidence with attention. I can imagine our friend de Man encountering a twentieth-century predecessor (less cellular disruption, not enthroned under the sink), having absorbed a similar set of murmurs, then coming up with those sentences streaming negations to parry every iconic turn with which it responds, weaponizes, conserves, relapses—relying on tropes of victimage, mourning, exclusion (until, as with the corporate person, that includes himself). Sentences such as the one cited above: "The most [*true mourning*] can do is allow for non-comprehension and enumerate non-anthropomorphic, non-elegiac, non-celebratory, non-lyrical, non-poetic, that is to say, prosaic, or, better, historical modes of language power." Feeding time in the Ideovomitorium. De Man's error was to pretend he could starve it into submission, block retreat, compel abandonment of an entire form of tectonic inertia.

Sometimes bored, I taunt it. Something like: "So this is the whizz-kid that reshaped earth." Or: "And all your universals boiled down to trolling raids on other enclaves and hominid types, nerves, representational systems, bodies, ecographies, collectives—all with a little cinema in your head playing reruns." On occasion we lock eyes in silence, the moments I wait for. Rather than turn aside, he grins, then just starts mumbling again to my face, surrounded by wrappers and undigested treats: ... "mmm.... Others—the animal, mmm"... "the singularity"... "human rights" (belch).

Ah, life without Anthropos... *things* just won't be the same.

Bangkok, July 1, 2015

Chapter Two

What is the Anthropo-Political?

CLAIRE COLEBROOK

1. Things "We" Have Been Told About the Anthropocene

If one heard the narrative from elsewhere one might imagine it would proceed quite differently. There is a species, and part of that species decides initially that it is exceptional at a moral and rational level, but then declares—late in the day—that the species is better described neither as moral nor as rational, but as constituted by destruction. One might think the next chapter would see this species go through a phase of humility, and yet that is not how the story seems to be unfolding. "The Human" seems to have been given a new lease of life, with knowledge of its destructive power (for all its bleak and dire predictions) nevertheless seeming to offer *knowledge* (once again) a certain privilege. If things are *this serious* then we need to throw away relativism and humility and start thinking *for real.* It seems as though, then, that the Anthropocene is not just one more claim about nature but that it erases all sublimity: "nature" is not some idea that we must assume but never know; nature is so real and so present as to have the force to erase decades of social construction, relativism, contestation, anti-humanism and *theory* (if theory is, as de Man claimed, that which is constitutively resisted precisely because to think theoretically would be to remain suspended before parsing matters into narrative form). Nature, now, offers its own narrative and frames the human species, placing it within the scale and register of earth system science.

Things "We" Have Been Told About the Anthropocene: first, we have passed a tipping point; second, this geological event is a game-changer and spells death both for social constructivism and post-humanism; and, finally, we are faced with a stark choice, either we radically transform the world or we are doomed. The Anthropocene is not only a geological declaration, but a series of diagnoses that generate imperatives. If we have damaged the planet to this degree, then we all need to act as one, and for the sake of saving us all. If we have passed a tipping point, then we are faced with necessary and immediate measures that cannot indulge in the luxuries of democracy or doubt. If there is a way that we might survive then that is the path we must follow. And who is the "we" that is being saved? It is the "we" that is constituted precisely by way of a death sentence: I mourn my future non-being and therefore I am. Further, I mourn my future non-being and therefore I must do all I can to survive. There is no longer time for post-humanism, anti-humanism, and—most of all—no time at all for questioning the prima facie value and existence of the human. The Anthropocene requires that we think of humans as a species, and (even though that unifying thought has been generated by evidence of destruction), once the species comes into being as a geological force its survival is constituted as an imperative.

It is in this respect that the Anthropocene is not one twilight concept among others, arriving late in the day to alert "us" to our destructiveness, and thereby demanding that we become the properly earth-nurturing humans that we ought to have been all along. If the Anthropocene—today—is possible, this is because its potentiality haunted the very constitution of the human as a political animal. How is it that "man" is thought of as *necessarily political* (as properly oriented towards others, as having his essential being defined by having no essence other than sociality)? To be *political* is not just to be, but to be *in-relation.* It is this fetish of unity, connectedness, attunement, mindfulness and humanity as an intertwined ecology that pervades post-Anthropocene pop culture (ranging from the utopian dreams of James Cameron's *Avatar,* to more recent sci-fi fantasies of collective consciousness, such as the recent Netflix series *Sense8,* or the earlier Fox series *Touch*). The notion of the world as a living system is not confined to earth science but seems to operate as a default mode of pan-psychism that—in turn—enables any number of narratives that overcome human and Cartesian atomism to find ourselves, our proper humanity, in a world that has found itself as what it ought to have been all along, truly earthbound.

This shrill insistence on *the political*—that there is, beneath it all a "we" that must emerge once we erase the pernicious "man" of capitalist individualism—allows for a humanity to come. (I would distinguish this from the Deleuze and Guattari conception of a "people to come," precisely because their use of the term follows from an erasure of a humanity that never was; for them, "man" was always a limited effect of certain and localized knowledge and desire formations.) Both in social theory and in post-apocalyptic culture capitalist "man"—the isolated individual of self-interest, consumption and myopia—is displaced by the new humanity brought into being by the Anthropocene: humanity is (finally) revealed as an aspect of the earth as a living inter-connected system. Perhaps this is what *post*-apocalyptic means: after the realization that what called itself "man" was destructive at a geological level it became time for a new eco-friendly humanity to inherit the earth.

One of the constant objections to the overly inclusive conception of the Anthropocene, has been that it is insufficiently political.[35] To refer to "anthropos" is to forgo asking just who within humanity was the agent of destruction. To politicize the Anthropocene is to place a broad geological narrative within a different scale or frame, and then to differentiate among humans. If it is capitalism, patriarchy, corporatism or colonialism that is responsible for geological inscription, then an observation regarding the species becomes nuanced by reference to a timeframe of a smaller scale. And this shift of scale becomes possible if one maintains a conception of politics that allows for significant groupings: rather than blame "humanity" we might think (as Klein does) of "capitalism versus the climate."

One might say, then, that the problem with the concept of the Anthropocene is not that it assumes that there is such a thing as humanity in general, but rather that it is insufficiently humanized, and that one would want to locate the "anthropos" within human historical narratives that make sense of what "we" do by some broader reference to relations among individuals (Moore 2014). To politicize is to offer a narrative, with narrative always generating a moral decision regarding scale. Debates about the "Golden spike," or debates regarding the temporality of anthropogenic destruction, like the Anthropocene in general, are claims made for narrative frames and trajectories, and presuppose not only that humans are political—defined by their relations to each other—but that those relations can be morally differentiated (Lewis and Maslin 2015). For all their complexity, the majority of these human-human relations of politics are relations of good and evil, and allow for the thought of a proper humanity that would *not* be guilty of the Anthropocene scar.[36] One might say that it is only when a crime has been committed, such as planetary destruction, that there becomes both the need to attribute that crime to a perpetrator, and then have those to whom the crime is committed emerge as the proper inheritors of the earth.

If one objects to the notion of the Anthropocene that it generates a far too general and all-encompassing "Anthropos" it does *not* follow that one needs to find another culprit or proper name, such as the Capitalocene or the Corporatocene. Other geological markers, including the Holocene, do not have a cause but mark a shift in multiple factors and forces; and perhaps one could think of all history this way, as ongoing reconfiguration with multiple acts of violence and opportunism. It is not as though there are masters who win and who inflict violence; there is violence and cruelty, from which something like a distribution between master and slave emerges. Only by way of narrative metalepsis could one think of a certain type of humanity (capitalist man) *causing* the Anthropocene; rather, it is from the observation of changes to the earth as a living system that one can then, from a series of observed patterns of violence, posit a relatively stable force or "anthropos." And only then, once that "man" as agent of destruction is posited, might one then find *another* agent, a force for good.

The briefest of glimpses at contemporary cultural production testifies to this *ressentiment*: if humanity has been the victim of those who have intensified its improper capacities (over-consumption, over-production, exploitation) then another humanity will emerge after the game-change of the Anthropocene. The flourishing industry of cli-fi and post-apocalyptic drama– in addition to all the usual laments regarding capitalism, colonialism and patriarchy—have generated an excess of monstrous humans, including aliens, zombies, viruses, pseudo-humans and corporations who enslave the planet, thereby prompting the proper future-oriented humanity to gather forces and triumph over an external and inhuman force. (One can think here of everything from *Avatar* of 2009, to *World War Z* (2013), *Oblivion* (2013), *Interstellar* (2014), *Elysium* (2013), *Into the Storm* (2014) and *Mad Max 4: Fury Road* (2015).) In both fiction explicitly concerned with climate change to more allegorical presentations of species bifurcation, humanity is at war with the improper and inhuman fragment of itself: a destructive humanity becomes the catalyst for human triumph, with a proper humanity emerging with sublimity from near death. (In *Interstellar, Elysium* and *Avatar* a myopic, profit-driven, corporate, militaristic, merely surviving State power is vanquished by an ecological, future-oriented and empathetic humanity." If there is an improper destructive humanity (and this is evidenced by the geological inscription of the "anthropos") then there *must be* a humanity who would emerge when such an evil humanity has been vanquished. If we are fallen, now in a world of loss and mourning, then there must have been (and will be) a better humanity to come. This is *not* to say that there has not been violence and injustice, but it *is* to suggest that just because there has been violence one might attribute such force to a single guilty agent (bad Anthropocene man) thereby promising another humanity. One might contrast a Kantian conception of justice, where the very idea of a good will (defined against the contamination of the present) necessarily opens and promises the idea of justice to come, with a Nietzchean notion of justice as *dike*: as the play of forces that generates disequilibrium. The former conception of justice allows the promise of a proper future to follow necessarily from the violence of the present: if there is no paradise, then paradise must have been lost, and therefore *will* be regarined. This logic is not confined to post-apocalytpic cinema, but is announced in Jacques Derrida's deconstruction, where despite all the violence undertaken in the name of justice and democracy, these ideas *cannot* be contained by the present and necessarily promise a future, justice and democracy to come. Against this, one might think of forces in strife as operating less by way of good and evil, or light and dark, and more by way of twilight—of discernible distinctions but always amid a potentially overwhelming indifference.

It does not follow, then, that all the evidence of violence and injustice, and especially the positing of an agent of destruction that operates at species-level, generates the promise of a good human future. If humanity has somehow managed to bring itself to the brink of non-existence, it does not follow that it must rally to save itself, nor that it must do so against a certain evil tendency that will be vanquished in the humanity to come, nor that "we" will be all the greater for having contemplated the potential end of humanity as such. From Naomi Klein's claim that climate change is the opportunity finally to triumph over capitalism, to the environmental humanities movement that spurns decades of "textualist" theory in order to regain nature and life, to wise geo-engineers who operate from the imperative that if we are to survive we must act immediately and unilaterally, the end of man has generated a thousand tiny industries of new dawns.

All of these vivid calls to arms rely on expertise and generate the very "we" that is being addressed. Let us take the first claim about Anthopocene tipping points, or the "golden spike" (Lewis and Maslin 2015). Let us accept the premise of the dispute, and say that we might quibble about just when to mark the Anthropocene (industrialized agriculture, colonialism, the steam engine, nuclear energy); accepting that there is a point where man became definitively destructive implicitly generates another pre-Anthropocene humanity, or a counter-Anthropocene. An implicitly moral line of time is effected: if there is a point at which humanity becomes catastrophic at a planetary level, then there is the possibility both of attributing blame, and of retrieving and saving another humanity.

A new humanity is constituted by the threat of its disappearance; or, to follow Nietzsche: it is the voice of a moral law ("Thou Shalt Not…") that produces "man" as a guilty animal, bred and groomed through the attribution of guilt. If there is a being called man who has destroyed the planet, then not only is a bad humanity produced as the new agent of history, an entire industry of those who would self-diagnose and redeem humanity becomes possible. It is as though only with the impending end of humanity does something like "the human" become visible in all its anti-human glory. Now that geologists have discerned evidence of damage at species level, the human becomes at once victim, agent and redeemer. The "anthropos" is produced through an event of guilt and diagnosis; if there is damage and inscription at a geological level, then there must be a response at global level, an end to all talk of there being no such thing as "man," and some account as to who, when and how this tragedy occurred. The "anthropos" brings himself into being by way of a blinding discovery: it turns out, after all these years, that there is a "we" and that "we" have not been good to the planet (well, at least not the planet as we would like it to be). States of emergency seem to call for a suspension of the free reign of opinion along with the resurgence of authority. If the bankers and economists benefited from the 2008 financial crisis by declaring that time was running out and "we" have to act now, and *then* think about justice, then one might ask why climate science with its dire predictions has not been blessed with the same unquestioning obedience. Perhaps it's because of a failure of rhetorical flair: "we" can only hear those who offer a future to come, a promise of a blessed humanity that will rightly inherit the earth. Indeed, there is no "we" outside this rhetorical call to arms. So while all the declarations of authority would call for one kind of action worthy of the dire predictions of the present—a questioning of whether what has called itself humanity has a right to survive —the only "action" has been an insistence on a future *for us*. Just as the 2008 financial crisis somehow—how?—seemed to prompt a desperate effort to *save* the banks, a climate crisis seems to justify saving humanity, and yet without all the emergency measures that were taken to "save" the global economy.

And yet, as Nietzsche argued, these idols emerge in moments of waning and decline, or disappointment, even if they are—for all their late appearance—eternal:

> ...as far as sounding out idols is concerned, this time they are not just idols of our age but eternal idols, and they will be touched here with a hammer as with a tuning fork,—these are the oldest, most convinced, puffed-up, and fat-headed idols you will ever find (155).
>
> The disappointed one speaks.—I looked for great men, and all I could find were the apes of their ideals (Nietzsche 2005, 161).

It may well be that it is only when the species is at an end that it recognizes itself as a species, and becomes fully and self-righteously human only in the moment it is required to face its loss, a loss that—in turn—seems to grant it the imperative to survive at all costs. Man exists, and must be saved. He can only be recognized and saved in these last hours, when destruction has reached such a degree as to become evident. Some have presented this moment as a *felix culpa*: without the evident, readable and diagnosable destruction of the Anthropocene, we would not have realized who we are, and might have proceeded with capitalism, industrialism and ecological destruction without this wonderful wake-up call. Now man knows who he is, and that he can only be saved by himself. He must first accept that there is indeed this unified global/geological being called the human. From there, in this moment of being too big to fail, all forms of emergency measures must be unfurled, if we are to survive. In short, it is by way of destruction that the human emerges, finally, as destroyer and preserver, enlivened by a whole series of moral laments that produce man *as he might have been*—the man prior to whatever we determined the golden spike to be—and then further enlivened by a new managerialism that accepts that if the human exists as destroyer then there is some urgent imperative to generate a fully human future. It is by way of a whole series of self-accusations that Anthropocene man becomes capitalist man, patriarchal man, corporate man, colonizing man, or the man of the nuclear age; this industry of self-accusation allows for another humanity, and one—precisely because it is threatened—that deserves to be saved.

Here, though, it might be worth questioning whether the Anthropocene is an event that really does demand that one either accept the general condemnation of man or blame a specific modality in man's history (capitalism, corporations, males, the West). It seems that we have two options: either the Anthropocene is an effect of man in general, or it can be attributed to capitalism (or corporations, or colonialism, or patriarchy), in which case man can emerge as an innocent animal—as a new humanity to come. But what if one were to refuse both these options by suggesting that man is neither the global culprit, nor the global victim, and that there are many living beings on this planet who live, dwell, struggle and survive with no sense of humanity in general? What would the present begin to look like if we refused both the claim for humanity as global agent and humanity as proper potentiality who may (and ought) to inherit the earth? If we accept the Anthropocene premise that man in general is responsible, then we accept something like the human as such and ignore the subtleties of history, culture and difference; if one aspect of man is responsible—say, capitalism—then that allows for a space outside the guilty party. Either way, one generates the human, first by way of accusation—the Anthropocene, a single scar that calls us all in the moment of defeat—or by way of exculpation: no, not the human in general, but these humans—the capitalists (say), whose end will actually allow us to dream of a new beginning. Those who declare man to be guilty are the first true humanists, generating the "Anthropos" as agent, and promising another humanity—one who can be intimated after the crime of ecological destruction has been detected, diagnosed and managed.

2. Theory Refuge

If the Anthropocene has seemed to erase decades of "theory" that waged war on the very conception of the human, then at least—we might console ourselves—this gives the humanities a clear task at hand. What is presented as a scientific imperative where man is now discovered as both agent of destruction and as endangered species, requires some new industry of reading and critique that will not allow the future to fall into the hands of corporate opportunists or those whom Clive Hamilton has referred to as the Prometheans:

> The direction in which we thought we were going has now been denied to us. The historical force of this should not be missed, for it means that the utopian promise of all political and religious ideologies, both materialist and metaphysical, vanishes. So we have reached the point in history where we must face up to the tragic consequences of 'the American way of life', a way of life also lived in other affluent countries, albeit typically with less intensity and ideological conviction. The same qualities that made the United States a great nation—relentless optimism, commitment to know-how, determination to expand—have become the enemies of its preservation and, collaterally, the preservation of the rest of humanity. A nation that has expansion running in its blood can barely

> conceive of contraction, and so the question we will soon be forced to ask is how much of the rest of the world will be sacrificed to prolong the dream of affluence? We have seen that it is not true that Prometheans must favour climate engineering and Soterians must oppose it. Nor is it true that Soterians are against technology. It is not so simple. Yet among those who believe we should make preparations to engineer the climate there is a sharp division between Prometheans and Soterians. The former are inclined to see it as a way of defending the established order so that expansion can continue uninterrupted. The latter see it as a regrettable measure to protect those deeper values now threatened by the consequences of endless expansion—viable societies, vulnerable communities, ecological values and life itself (Hamilton 2013a, 208).

Hamilton proposes that we abandon the social sciences (those who insisted that the human was a mere construction), and should temper technocracy with questions of politics. After years of theory that contested every naturalization of what was ultimately historical and political, "man" has returned. Feminism had tirelessly criticized not only the notion of a generic "man," but had also increasingly questioned the category of "woman." If, as Catherine Belsey argued, "bourgeois liberal humanism," was a blindness to the ways in which the subject or "man" had been composed from contingent, and therefore alterable, political relations, then theory was necessarily the demystification of anything so naively timeless as the individual: "The ideology of liberal humanism assumes a world of non-contradictory (and therefore fundamentally unalterable) individuals whose unfettered consciousness is the origin of meaning, knowledge and action" (Belsey 1980, 56). Terry Eagleton, also, defended theory as an awareness of the historical and social forces that had been (ideologically) masked as "natural"; theory is always political and to say it is not so is just another political maneuver:

> There is, in fact, no need to drag politics into literary theory: as with South African sport, it has been there from the beginning. I mean by the political no more than the way we organize our social life together, and the power-relations which

> this involves ... Even in the act of fleeing modern ideologies, however, literary theory reveals its often unconscious complicity with them, betraying its elitism, sexism or individualism in the very 'aesthetic' or 'unpolitical' language it finds natural to use of the literary text. It assumes, in the main, that at the centre of the world is the contemplative individual self, bowed over its book, striving to gain touch with experience, truth, reality, history or tradition (Eagleton 1996, 170-71).

Eagleton's work is typical of theory *as politicization*, as an unmasking or de-naturalizing of "the human." The putative bourgeois liberal subject who has no qualities other than the capacity to be open and tolerant of all others who (in turn) have no relevant qualities other than their own formal capacities for reason is nothing more than a mask for the capitalist consumer; the liberal public sphere, or the level playing field of free public reason is just an alibi for the free circulation of capital. Differences of race, sex, class, gender, ethnicity, ability and even the personhood of nonhumans have been insisted upon against the undifferentiated and depoliticizing blandness of humanism. The supposedly universal "human" was always white, Western, modern, able-bodied and heterosexual man; the "subject" who is nothing other than a capacity for self-differentiation and self-constitution is the self of market capitalism. To return to "anthropos," now, after all these years of difference seems to erase all the work in postcolonialism that had declared enlightenment "man," to be a fiction that allowed all the world to be "white like me," and all the work in feminism that exposed the man and subject of reason as he who cannibalizes all others and remakes them in his image. The Anthropocene seems to override vast amounts of critical work in queer theory, trans-animalities, posthumanism and disability theory that had destroyed the false essentialism of the human. The "human" of the late twentieth century had increasingly become a humanity of difference, defined less by being than an ongoing strategy or performance or becoming. But this humanity of becoming and self-differentiation was possible only by way of a negative universalism, where the human was unified by having no essence other than that which it gave itself through existence. The politics of this humanity was a politics of increasingly recognized difference, achieved by analyzing the dynamics that constitute the human in its specificity (rather than its being as species).

If one accepts this notion of the political—that one should resist false naturalizations or universals—then to politicize the Anthropocene becomes a question of determining what scale or scales would generate the proper frame for narrating the genesis of "man." Should one read a text in terms of social-economic narratives, or in terms of the founding event of sexual difference, or as an ongoing reinforcement of heteronormativity, or as a document of Western reason, imperialism or racism? Such questions are theoretical because they embed seemingly literary, personal or aesthetic texts within other contexts. Against a bland and undifferentiated humanism, political theorization strives to be more intensely and differentially human—seeing humans as emerging from complex political relations, where politics has to do with social formations and what is brought into being (historical/cultural) rather than what simply is.

We might pause here to note that theory has two tendencies. The first would generate theory as *theoria* in its classical philosophical sense: theory steps back from, or disengages from, relations, and does not (yet) have a world in all its richness of affect and meaning. Theory would be, as de Man suggested, an attention to *reading* rather than assuming that the matters of a text simply offered themselves as meaningful. Materiality, in de Man's sense, is very far from "historical materialism" (where matter is the matter of a history of human labour), and also very far from a whole series of "new materialisms" where matter is a "turn" towards bodies, affects and the vibrancy of matter. On the contrary, for de Man "materiality" suspended any turn back to whatever tropes had turned away from; to read was to attend to the matter of the text. In a different mode (but with similar force) Lacan would refer to the "agency of the letter"—the capacity for inscriptions, traces, marks or *literally* letters to operate with their own force. Perhaps this is why theory, at least in this mode, was deemed to be anti-political, guilty of abstraction and formalism. Many who objected to high theory objected to abstraction, textualism, depoliticization and a series of losses (of history, context, bodies). The second sense of theory reverses this conception of distance or disaffected looking and insists upon politics; to be theoretical is to historicize, contextualize, and be mindful of multiple identities, investments and experience.

When de Man wrote about the necessary resistance to theory—that reading would always lapse back into the affirmation of a meaning, *there,* in the text, to be unfolded—he appeared to be guilty of the most depoliticizing mode of theory, of assuming that texts were not truly grounded in real, material, violent, worldly and human matters. Our claim is not that de Man was, after all, properly political, but that his de-politicization did not go far enough. His attempt to free the matters of text from meaning and security nevertheless deployed a language of tropes and rhetoric that (at first glance) spoke of materiality in human linguistic terms, almost as if one could think about language as having an intentionality, or as if language were anthropomorphic. In de Man, though, there is already the de-politicizing sense that the anthropomorphism of rhetoric does not refer to humans using language to render the world in their own terms, but that the figure of "the human" is an effect of a failure to read. In this respect, we might say that the problem with the Anthropocene is not that it depoliticizes, by placing all humans as guilty agents, but that it is overly political: still assuming something like "man" whose human-human relations form polities (rather than seeing whatever emerges as human as an effect of what Nigel Clark terms "inhuman nature" [Clark 2011] or Timothy Morton refers to as "ecology without nature" [Morton 2007]).

There have been many death sentences, elegies, autopsies, mournings and celebrations of the death or end of theory. Far more forceful than those who mourn theory are those who maintain its afterlife by way of a redemption narrative. Robyn Wiegman has even provided a helpful account of the "reparative turn," which she applies to queer theory, but which is pertinent well beyond queer and gender studies (Wiegman 2014). Theory was once, supposedly, textualist and apolitical, but thankfully there were a series of realist, ethical, materialist, reparative and vital turns that restored theory to its properly political place. To theorize is to politicize, and to politicize is—ultimately—to situate whatever is inscribed within human agonistics. Theory—from Derrida's "ethical turn" to new materialisms and the insistence on vitality, vibrancy and life—is no longer the theory that de Man claimed was necessarily subject to resistance (the theory that takes any political and human reality and sees nothing more than inscription). After many post-theoretical *mea culpas*, reforms, and "turns," theory has become politics, and politics has become hyper-human. To read politically is to focus on the composition of the human, with composition implying malleability and a future that might be otherwise.

If theory has become an attention to construction and composition, the Anthropocene often appears as a reactionary insistence on the real and non-negotiable. Indeed, it often seems as though it is theory as such that seems to have fallen victim to the new scale of the Anthropocene. Or, "theory" has less to do with "theoria" or reading, and far more to do with trends in topics: from bodies and animals, to plants and now the Anthropocene. "Theory" has come to denote taking a stand, looking at the world from the standpoint of one's lived experience, identity or allegiance. If late twentieth-century theory was—ultimately—a politicization of all that appeared as self-evident and natural, and an overwhelming objection to anything as static as the human, then the Anthropocene (at least for some) corrects the notion that the human is a figure or construct within history. Against the "modern" diremption of humans from nature, Hamilton announces an end to social science and social constructivism (Hamilton 2013b). Humans are all too real, bound up as they are with the earth as a living system.

> ...the effects of human-induced warming go far beyond changes in the weather; everything is now in play, and not only scientifically. So let us now make a leap from the land of science to the grounds of the humanities, because the astounding new facts uncovered by Earth system science force us to rethink our understanding of history. The idea that humanity makes its own history and does so against the backdrop of the Earth's slow unconscious evolution is deeply implicated in modernity. We are accustomed to thinking of humans, having emerged from the primordial darkness, as independent entities living and acting on a separate physical world, a world we plough up, mine, build on and move over but which nevertheless has an independent existence and destiny. This understanding of the autonomy of humans from nature runs deep in modern thinking; we believe we are rational creatures, arisen

> from nature, but independent of its great unfolding processes (Hamilton 2013a, 194-95).

For Hamilton, the human is not a political, social or constructed force, but an aspect of an earth system that needs to be studied at a geological scale, which is—also—the properly political scale. Those who want to debate the starting point of the Anthropocene—say, with colonization in the seventeenth century are, Hamilton insists, stuck in environmental science and have not matured to earth system science, where the point is not relations among humans but the earth as a complex whole (Hamilton 2015). Politics that operates at the traditional level of the polity (by way of relations among humans) needs to be opened beyond the "modern" enclosure within human history: "From hereon our history will increasingly be dominated by 'natural processes', influenced by us but largely beyond our control. Our future has become entangled with that of the Earth's geological evolution. ...contrary to the modernist faith, it can no longer be maintained that humans make their own history, for the stage on which we make it has now entered into the play as a dynamic and capricious force" (Hamilton 2013a).

On the other hand, and running directly against the shift to the scale of humans as a species operating as geological agents, various counter-Anthropocenes have been declared. If we define this epoch of destruction as the Capitalocene or the Corporatocene (confined to a specific type or aspect of humanity, or a political grouping) then another world and another polity would possible, such as the "Sustainocene" (Manahan 2014), or the Gynecene:

> We declare the imperative necessity for a new geological era to be commenced, before the Anthropocene is even officially admitted on that scale (it might be that by the time it gets fully acknowledged, it will be too late). Rather than continue to contemplate our annihilation, contributing to it or declaring hopelessness in front of it, we should at least try another approach—and this approach has to exclude patriarchy in all its expressions and institutionalized forms of violence: domination, exploitation, slavery, colonialism, profit,

> exclusion, monarchy, oligarchy, mafia, religious wars (Pirici and Voinea 2015).

If the human is no longer accepted as a timeless political entity—the atomic subject of liberal theory—this is because he is bound up with complex dynamics of multiple strata; to substitute terms such as "Anthrobscene" for "Anthropocene" is not to deny the existence of the human; quite the contrary, it is to see the human as effected and effective at different and multiple scales. This is what has led Jussi Parikka to tie the earth, media ecologies, and what appears as the Anthropocene into an "alternative account" that would not allow geology to co-opt the narrative of our time:

> Temporalities such as deep time are understood in this alternative account as concretely linked to the nonhuman earth times of decay and renewal but also to the current anthropocene of the obscenities of the ecocrisis. Or to put it in one word: the anthrobscene (Parikka 2015, 64).

More explicitly concerned with theory's failure to deal with the complex relation among humans, life, objects and labor, McKenzie Wark declares the need for competing narratives and a shift in scale, where politics would no longer concern relations among humans solely, but would include the earth and technology:

> It's time for other stories. One thing, strangely, that the Soviet Union and its American Cold War nemesis had in common was a clear understanding that any narrative for a whole way of life has to scale. It had to work for millions, even billions. The challenge then is to take the desire for alternatives from the romantic anti-capitalists and wed it to the grand scale on which capitalist realism imagines itself now to be the only discourse (Wark 2015, 217-18).

For Wark theory can be saved by climate science, where climate does not refer to weather patterns or even ecology in its usual sense, but all the bodies, technologies and systems that compose a complex whole:

> Climate science has no need for Marxist theory, but Marxist theory has need of climate science. Its project has to move

> on, from the critique of political economy, to the critique of its Darwinian descendents in biology, to the critique of climate science, as a knowledge which shapes the most general worldview of the Anthropocene. It may sound like hubris to name an entire epoch the Anthropocene, as this seems at odds with the decentering and demoting of the human that was a significant achievement of ecological thought. But the Anthropocene draws attention to androgenic climate change as an unintended consequence of collective human labor.
>
> The Anthropocene calls attention not to the psychic unconscious or the political unconscious but to the infrastructural unconscious. What labor builds is a disingression. Moreover, viewed via the Anthropocene, human action remains quite modest and minor, it's just that in fluid systems a small action can have disproportionately large effects (Wark 2015, 180).

Wark's aim is to broaden politics beyond the human, and to broaden the human beyond man as a political animal; man is a laborer and machine, bound up with non-human laborers and machines. In short, he is a cyborg, but no less real and worthy of concern for all that. Here Wark draws upon Donna Haraway who has also questioned the timing of the label of "anthropos": after all the decades where it was unthinkable to do "good work" while assuming the human as the basic unit of analysis, we might be better off thinking in terms of the Capitalocene, Plantationocene, or Chthulucene (Haraway 2015). For Haraway, such questions of naming are questions of scale:

> ...the issues about naming relevant to the Anthropocene, Plantationocene, or Capitalocene have to do with scale, rate/speed, synchronicity, and complexity. The constant question when considering systemic phenomena has to be, when do changes in degree become changes in kind, and what are the effects of bioculturally, biotechnically, biopolitically, historically situated people (not Man) relative to, and combined with, the effects of other species assemblages and other biotic/abiotic forces? No species, not even our own arrogant

> one pretending to be good individuals in so-called modern Western scripts, acts alone; assemblages of organic species and of abiotic actors make history, the evolutionary kind and the other kinds too (Haraway 2015, 159).

Such an objection to the Anthropocene—man returning as a unified species—seems to commit the depoliticizing crime that Roland Barthes (so long ago) described as creating a static nature by way of freezing history:

> This myth of the human 'condition' rests on a very old mystification, which always consists in placing Nature at the bottom of History. Any classic humanism postulates that in scratching the history of men a little, the relativity of their institutions or the superficial diversity of their skins ... one very quickly reaches the solid rock of a universal human nature. Progressive humanism, on the contrary, must always remember to reverse the terms of this very old imposture, constantly to scour nature, its 'laws' and its 'limits' in order to discover History there, and at last to establish Nature itself as historical (Barthes 1972, 101).

Fredric Jameson sustained this demand in his imperative, "always historicize" (1981, 9). The history, here, that would destroy the undifferentiated "anthropos" is a history of colonialism, labor, sexual contracts, slavery, capitalism and corporations. The history of deep time or geological scales would need to be nuanced by smaller scales that differentiate "man" and that open spaces of refuge that would demarcate capitalist man, and find a space for his less guilty others. "Man" is not reduced to what he actually is, nor what has occurred historically, nor as he is now defined by geology and earth system science. Political reading would distance itself from the given, the seemingly natural and universal. To read or think politically is to refuse *the human,* thereby enabling us to say that where we are now is an outcome of hegemonic interests (one group of humanity, one portion of the species) and that there might be another humanity, another given, another composition. This is how theorists as abstract as Alain Badiou have defined the political, as an event that breaks with what appears to exhaust the real: "politics can be defined therein as an assault against the State, whatever the mode of that assault might be, peaceful or violent.

It 'suffices' for such an assault to mobilize the singular multiples against the normal multiples by arguing that excrescence is intolerable" (Badiou 2007, 110). Far less abstractly, and in direct response to the declaration of the Anthropocene, Bruno Latour has insisted on the composition of the world as it is at present, and therefore the possibility of its transformation, its capacity for being re-made:

> There is nothing native, aboriginal, eternal, natural, transcendent in the habits that have been framed during the few centuries "market organizations" have exercised their global reach. No feature of *Homo oeconomicus* is very old: its subjectivity, its calculative skills, its cognitive abilities, its sets of passions and interests are recent historical creations just as much as the "goods" they are supposed to buy, to sell and to enjoy, and just as much as the vast urban and industrial infrastructure in which they have learned to survive. What has been made so quickly can be unmade just as quickly. What has been designed may be redesigned. There is not fate in the vast landscape of inequalities we associate with the economy and their unequal distribution of "goods" and "bads", only a slowly built set of irreversibilities. Now that historicity has shifted from the stage to the backstage of human action — namely, from second to first nature — activists should ally themselves with the globe against the global (Latour 2014a, 12).

If the Anthropocene is really the "Capitalocene," or the "Corporatocene," and if it can be identified with high industrial appropriation, and colonialism then there can be another humanity that would not be guilty. We might, as Naomi Klein, has done declare climate change to be the opportunity for us, the humanity who objects to capitalism, to find justice.

3. What is the Political?

When the Anthropocene emerged as a concept it seemed to offer, to me at least, so much promise: and by promise, I suppose, I mean something quite different from an almost automatic promissory response in humanities inquiry. When Jacques Derrida wrote on the exorbitant nature of some of the humanities' key concepts—such as justice, democracy and promising—he generated the possibility of a profound break with the temporality of concepts. To a great extent concepts became, by way of Derrida's work, *essentially* promissory. If there were to be such a thing as justice, democracy, forgiveness, hospitality or promising then the very limited and impoverished nature of actuality was itself a guarantee of the infinite. Justice could not be reduced to any of its actual instances; if one uses a concept then one relies upon it being recognized because it exceeds any single context. Language is essentially in excess of any single instance of use. This means both that there can be no such thing as good conscience (for any ethical exercise of responsibility must exclude all the other potential fulfillments) and also that there can be no exhaustion of ethical concepts: justice, forgiveness, democracy and friendship have a power—as repeatable concepts—to extend, infinitely, into the future. Here, Derrida's ethics operates by a structure that is akin to Kant's critique of pure reason. Derrida's ethics relies on what can be thought but not given, with *thought* being opened to the infinite by inscription: every concept "I" use has a power to be repeated beyond any use. One cannot contain or limit a concept on the basis of use or intention. Kant's critique of reason accepts what, to use Derridean language, we might refer to as necessary impossibility: the world we know and experience is (by virtue of the fact that it is *known*) different from ourselves. We therefore know and experience in relation, and this relation to what we are not is only possible because we have a space in which we are located; to be in a relation to something or to know something, is to be different, distant, deferred. We experience a difference between ourselves and what we are not, and this necessary space that enables the experience of difference between knower and known also requires time. If we knew everything all at once, in a fully self-present and divine intuition without difference, delay, or deferral of a further revealed world, we might not call this knowledge at all but simply absolute self-presence outside all time and space. Time and space do not limit but enable knowledge, but if this is so then knowledge is essentially bound up with difference and deferral; it is never absolute. It follows for Kant that we should not lament the distance we have to the world that would supposedly prevent us from knowing: rather, this distance and difference is the condition for knowing. Because we know things in relation, and in time, we keep seeking further experience and knowledge, and this leads us to imagine or strive for a knowledge without distance—we cannot erase that tendency, but we cannot fulfill it either. Kant has his own ethical theory based on practical reason, where we imagine ourselves *as if* we were not bound by such finitude (as legislating universally, freed of all self-interest). Derrida's ethics does not indulge in such a supra-human thought of a pure will that would *think* as if it were not bound up with the determination of the world. On the contrary, as I have suggested, Derrida follows an acceptance of the limits of our knowing and conceptualizing, and this itself yields an essentially promissory ethics:

> The necessary disjointure, the de-totalizing condition of justice, is indeed here that of the present—and by the same token the very condition of the present and of the presence of the present. This is where deconstruction would always begin to take shape as the thinking of the gift and of undeconstructible justice, the undeconstructible condition of any deconstruction, to be sure, but a condition that is itself in deconstruction and remains, and must remain (that is the injunction) in the disjointure of the *Un-Fug*. Otherwise it rests on the good conscience of having done one's duty, it loses the chance of the future, of the promise or the appeal, of the desire also (that is its "own" possibility), of this desert-like messianism (without content and without identifiable messiah), of this also *abyssal* desert (Derrida 2006, 33).

No decision I make can determine what counts as justice, but far from falling into what Gilles Deleuze described as a highly bourgeois ethics of compromise ("on the one hand.... on the other hand" [Deleuze 1994, 225]), Derrida used the failure of determination to insist that our thoughts of what might count as justice, hospitality, democracy, friendship or forgiveness would always be able to open to a future that was essentially impossible. For Derrida, then, it is the *violence* of the law (or its inscription that must emerge in a manner that is groundless and without precedent) that also, necessarily opens to a future of justice that is unconditional. He writes of an "originary performativity":

> … that does not conform to preexisting conventions, unlike all the performatives analyzed by the theoreticians of speech acts, whose force of rupture produces the institution or the constitution, the law itself, which is to say also the meaning that appears to, that ought to, or that appears to have to guarantee it in return. Violence of the law before the law and before meaning, violence that interrupts time, disarticulates it, dislodges it, displaces it out of its natural lodging: "out of joint." It is there that *différance,* if it remains irreducible, irreducibly required by the spacing of any promise and by the future-to-come that comes to open it, does not mean only (as some people have too often believed and so naively) deferral, lateness, delay, postponement. In the incoercible differance the here-now unfurls. Without lateness, without delay, but without presence, it is the precipitation of an absolute Singularity, Singular because differing, precisely [*justement*], and always other, binding itself necessarily to the form of the instant, in imminence and in urgency: even if it moves toward what remains to come, there is the pledge [*gage*] (promise, engagement, injunction and response to the injunction, and so forth). The pledge is given here and now, even before, perhaps, a decision confirms it. It thus responds without delay to the demand of justice. The latter by definition is impatient, uncompromising, and unconditional (Derrida 2006, 37).

Our inability to imagine complete or final justice should not yield resignation, nor—as we start to think about climate change—should we accept a cost-benefit analysis of how much justice we can afford; no conception of justice and its possibilities can forestall what justice as a concept might become, might do, or might offer. Here, then, is the promissory force of Derridean ethics: we live in a world of forces, relations, structures, inscriptions, and potentials *not* of our own making. We might claim with scientific certainty that justice is not possible; but if that certainty is truly scientific, it must accept the possibility of further experience and revision. The lack of absolute knowledge is neither paralyzing (because waiting until absolute knowledge arrived would be to forget the relational nature of knowing) nor compromising, precisely because every decision made in the absence of knowing bets on a future that (for all we know) may indeed come into being. Nothing can be said, here and now, to be impossible or to be closed down or determined once and for all. The very existence of concepts such as justice, democracy and hospitality enables the promise of something beyond all conceived present possibilities: the only impossibility is the determination in advance that certain events would be impossible.

When the Anthropocene was first articulated there had already been some anticipatory criticism of the arrow of time in the Derridean future: it seemed to go only in one direction. Because we exist within differences and forces that have a power and potentiality beyond the present, and beyond the power to determine the present, nothing can be fully determined, and no sense of a concept—of what a concept might do—can be saturated. Even the past we carry with us (such as the archive of philosophy, literature, science and inscription more broadly) has a capacity to "live on" in the absence of any of its original intentions or forces. This *mal d'archive,* far from prompting us to fall into despair and nihilism—that a text can mean just anything, and that there is no truth—generates quite the opposite: the text has a force that exceeds the limits we wish to place upon it, and its feverish, anarchic and untamed capacities have a generative power beyond any already formed instance. We might, following Derrida, say that everything has promise, or operates as a promise: to speak, to experience, to act, to write and even to destroy anticipates a future that is *not yet,* or that is "to come." Everyday experience of promising knows this to be true: from promises to pay back debts to promises to love or deliver justice there is always a gap between what we fully intend and give all our passion to achieving, and a life that can disturb, disrupt and erase the context that enabled the promise's bind. Everyday experience knows *also* that the promissory arrow of time is sometimes not an arrow into the future, sent by the present, but comes from the future.

Well before the Anthropocene was posed as a way of thinking about what would be readable in the future, climate scientists had been offering dire predictions. What makes a prediction differ from a promise? A promise—even if it cannot be commanded by intentionality or an anticipated future—makes some claim to retaining or realizing into the future. Phenomenology had argued that in order to experience something as present, as real, or as having being, I retain the past (retention) and anticipate future coherent experiences (protention); all experience is promissory. Even if I assume that something is *not* real, or is illusory, it is because I anticipate that its image will disappear on further examination. Most of Derrida's work on truth and ethics is of this anticipatory and promissory register: in order to have a science of geometry I experience in the here and now what *would be true* for a presupposed humanity (Derrida 1978a). That capacity to have an ongoing truth or ethics of the future — such as justice to come—*does* retain an archival past that may or may not be retained, but the future is both necessarily anticipated, and necessarily irreducible to protention and anticipation.

Predictions have a slightly, but importantly different, temporal structure from promises. They rely not on "us" retaining and anticipating; they happen *to us, from elsewhere.* To predict climate change or to say that changes that have already occurred are because of human actions that will continue to have effects beyond our intentions, is to do something quite different from promising. It is, in many ways, an exposure of a counter-intentionality of the promise: in addition to wanting to promise justice, love and prosperity (and doing all I can to bring these about) there is an ethical dimension of calculating the chances of fulfilling my promise. Promising is beautiful, and creative—it opens up to a future that is not yet. Dire predictions are ugly, and they are ugly because they make promising less easy. Dire predictions make the essentially ethical—the impossibility of foreclosing justice—something that we can still imagine and even strive for, but something whose striving and promising is bound up with what is promised not by us but *to us*. It is as though the future—even though it too lacks full authority and determining power—is making a promise to us. This future may not come about, just as promises may not be fulfilled, but the positive futurity of promissory time, is necessarily intertwined with a future that is—in predictions—promising us with a dimension beyond our potentiality and its imagination of impossibility.

What the Anthropocene adds to climate change prediction is both a difference in degree, and a difference in kind. By posing that humans will be readable as a geological force the Anthropocene seems to be destroying once and for all the future arrow of promissory time. By suggesting geological impact, and not just change within the human milieu, human time—the time of justice and the polity that had been increasingly disturbed by climate change—is now displaced. Other temporalities and scales come into play. We are perhaps today far less likely to be just to future humans and future life. Or, put more accurately, it is far less likely that there will be a time in which we will have been just. From a strictly Derridean point of view, far from us therefore abandoning all promise to the future, the stakes of justice would be raised: because we cannot definitively erase the possibility of future justice, because no event can discount the potentiality of justice (including the direst predictions), the future seems to provoke us not to close down possibility. By the same token, and still within the realm of Derridean possibility (though perhaps not quite the spirit of Derrida's thought), remaining within what the present enables us to promise should not foreclose reading what the future is promising us.

What the future promises us, at least from one mode of the Anthropocene, is the impossibility of promise: this is not definitely the case, for it is not certain that promising will be impossible, but it is certainly one possibility. Put in concrete terms then, I want to suggest that the Anthropocene has two modalities and they are both perversely sublime or counter-sublime. I want to call the first Anthropocene sublime, following Paul de Man, the recuperative sublime, and then gesture towards a second, material sublime. It is possible to say that the Anthropocene offers, by way of geological inscription a time-frame beyond our present, and that in doing so it opens human thinking to a future not given. What appears as a dire prediction—the earth as a living system marked with destruction—is also, by virtue of opening a time of thought beyond ourselves, the recuperation of the infinite. Some matters—geological inscription—open the thought, anticipation or promise of the "not yet." In the very mark of our defeat and limit we are given a time to come; we are given a "we", a "humanity to come." As I have already suggested, this can occur in the banalities of pop culture's cli-fi productions, where the presentation of humanity's end generates a heroic humanity that finds itself triumphant precisely by hailing its nobility in having been defeated, but also in ostensibly more high-brow moments, such as Quentin Meillassoux's insistence that God (as true redemption) must exist as a possibility precisely because of the injustice of the world at present. If the only thing that we can posit as absolute is contingency, then God must at least be possible; whatever is might be otherwise: "From this point on, God must be thought as *the contingent, but eternally possible, effect of a Chaos unsubordinated to any law*" (2008, 274).

By contrast de Man's material, non-recuperative, tropological, sublime would not allow some matters—such as geological inscription, or the thinkable contingency of anything at all—to promise or open anything other register. Such a refusal of reference, a refusal to see "the" Anthropocene as the definitive figure of "the" human might be referred to as sublime precisely because the lapse back into reference, meaning and recuperation is impossible to avoid. Even so, to approach what has been offered as the Anthropocene in terms of sublime materiality, would be to read geological inscription *as it is,* and *not* as some promise or sign of what humanity must do, or what humanity must have been. If politics has become the promise and necessity of another world *for us,* I would suggest that what the Anthropocene promises is an impolitic erasure or deadening of those matters, inscriptions, figures and substitutions that seem to stand for a world to come.

What does it mean to politicize a text? From one point of view—albeit a highly parochial one—to politicize a text is to offer a theoretical reading: (finding the right scale, or semantics). That, at least, was how Terry Eagleton (1996) made a claim for theory as such. If you simply say that a text *is* about 'x' (whether that be the English civil war or evil) then you are not accounting for how that relation between text and referent has been generated. You have a theory, a politics or an ideology (but you say you don't). Theory steps back from saying what a text means, and then asks how it means: to account for such relations, *non-politically* (by reference to what the author or text itself simply said) is the de-politicizing (and therefore political gesture) par excellence. Your scale is the individual or, worse, the face; the cult of authorship that ties writing to sentiment, authenticity and psychology. This was why Lukács objected not to certain readings of modernism, but to modernism per se with its stream-of-consciousness and presentation of the world as so much psychological flux and affect (Lukács 1963). To talk of feeling or experience, spirit or inwardness—these were (once upon a time) *the* hallmark gestures of depoliticization. Such experiential or "humanist" readings were deemed to be ideological because they took what was politically (i.e. historically and economically) constituted as natural or as personal. To say that the personal is political, or to say that what appears as "nature" is reified history, are gestures of shifting and expanding scale. To think of the individual or experience as the basic unit of analysis was, for theory of the late twentieth century, to occlude the socio-cultural formation of selves and ways of seeing and feeling. The problem of the affective fallacy increasingly came to be a political problem: to be talking about psychological rather than political realities. Today things have shifted, and affect is now the political concept of the day: but this affective turn is still both a shift of scale—from the self or polity to its smaller components—and a way of thinking *against* the ideology of the human as a coherent, universal subject—for it is affect, now, rather than history that provides an "outside" to what appears as natural and unalterable (Protevi 2009).

If ideology is not false consciousness (seeing what is as what it is not), and not simply error that might be corrected, then it is because it is constitutive and productive, or (to borrow from Fredric Jameson, borrowing from Lacan): ideology is the imaginary way in which individuals live their relation to the real (Jameson 1977). If we take this in one sense then we might see ideology as a rendering coherent, or tolerable, of matters that would otherwise be too traumatic to be lived. And this is how Jameson follows Adorno in thinking of ideology as the way in which the complex shudder of existence, contingency, and the violence of damaged life are rendered into some Manichean form of otherness: the nightmare of capital can be figured in forms of hero-villain narratives, or the sheer contingency of imposed systems can be figured by way of a wholes series of "natural" oppositions. For Jameson, to say that the Imaginary is the way individuals live the relation between the Real and the Symbolic, is to point to ideology as naturalizing and rendering into narrative form the cruel political oppositions of a history of class antagonism.

The Lacanian inflection of this phrase changes things somewhat, and brings us closer to where I want to go with de Man's notions of defacement and disfigurement. If we think of the Imaginary as the way in which individuals live the relation between the Symbolic and the Real, then we are shifted well and truly away from false consciousness, and from ideology as a symbolic rendering of political intolerability. For Lacan, it is the notion of the Symbolic as having a relation to the real that is Imaginary, *and* that constitutes me as a subject. We might say that what calls itself human *is* just this illusion or trope: that some matters stand for, present or enable the true experience of others. This is what de Man referred to as phenomenalization. To correct someone for being ideological or apolitical—to say that we need a *political reading*—is usually to shift scale, and to insist that what we see as "x" is really "y": what you take to be natural, I take to be political. The substitution of "political" for "natural"—*no, this is not natural; this is the outcome of history, human agonistic history*– keeps a certain figure of nature and politics in its place: nature is re-formed as that which is subject to human history, meaning, labor and transformation. Problems that appear as existential, abstract, universal, "natural" or personal are (depending on one's theory) deemed to be an outcome of familial, colonial, class or economic dynamics. And, if we accept Jameson's notion of ideology, all dynamics come down to historical dynamics—and human historical dynamics at that: what looks like racism or a loathing of the other is the way in which ultimately political (economic) conflicts are rendered tolerable. One of the problems in accusing a reading or a text of being ideological, a-political or of requiring *political* analysis is the way in which shifts of scale are often figured in terms of extensive space and size. You are ideological because you take a relation to be personal (the relation between two individuals) when you need to see it as a fragment of European history or patriarchy. It is for this reason that seemingly temporal shifts in scale tend to be mapped by way of a certain figuration of the globe (and not that of the planet). Jameson's history, for example, is the history of capital, a history not only of a certain scale, but also of a certain space.

Let us take a standard case of shifting from a depoliticized or metaphysical reading to a political reading. I might read Joseph Conrad's *Heart of Darkness* and claim that it is a novel about alterity, about the relation between human order and radical contingency, and *then* (in the narrative framing of the tale) about the ways in which this fundamental and transcendental conflict, this encounter with "what resists symbolization" is figured in terms of light and dark, and is rendered into a narrative). Or, similarly, I might read Albert Camus's *L'Etranger* as a text about the authenticity of a decision: faced with a world of constituted laws, norms and social coherence, Mersault acts and then refuses to justify his act—perhaps just as Kurtz in *Heart of Darkness* "cuts himself loose from the earth." Now, against this notion that texts are about something as transcendental or universal as evil, otherness or contingency we might want to say that such abstractions are there to make sense of specific political conflicts: what is really going on in *Heart of Darkness* is a localized, historical, institutional and contextual event of power: the imperial exploitation of Africa and its resources, a thoroughly political opposition that cannot be justifiably presented without some grand narrative of civilization. The text, therefore gives symbolic form (light/dark, past/ present, primitive/civilized) to what needs to be understood ultimately as an event of *material* conquest; by material we would mean not only the matter of this earth, but also a history of commodities, acquisitions, trade routes, enslavements and a racism (or horror of the other) that is required because the notion of power achieved by seizure rather than merit is not too pretty if you look too closely. We cannot read *Heart of Darkness* outside a more specific narrative of Belgian and British colonialism, and we cannot read *L'Etranger* without a sense of twentieth-century France's relation to Algeria and various *others* who are first rendered other by way of conquest and acquisition, and then metaphysically, mystically and ideologically rendered as other in some profound existential manner. The political, then, shifts from abstract opposition (some profound but unfathomable, and unmastered, and unsymbolizable relation (or *non-relation*) and insists that what is presented as mystically contingent and ungraspable is the outcome of human historical forces, a world mapped according to conquered, acquired, expropriated and stolen forces. A reading or "understanding" is ideological if it mystifies rather than narrates, if it presents as simply oppositional what is the outcome of history (by which we mean, who did what to whom).

As a recent occasion I might cite a *New York Times* opinion piece (now expanded and published as a book) by a young philosopher who had completed a tour of duty in Iraq. Philosophy, he pointed out, is about learning how to die, about being able to shift our attachment and attentions from pathological self-interest to a purview beyond our anthropomorphic narcissism (Scranton 2013). To this he was met with a stern correction from the left: the Anthropocene is *not* the Anthropocene but the corporo-cene; it was rapacious capital (not the suffering humans of the Maldives) who wrought this destruction, and it is corporations who should be held to account (Rugh 2013). It is not resignation in all its existential and tragic pathos but resistance that we need to affirm. Philosophy should not be about learning how to die, but learning how to change a world in which it seems that all we can do is die.

De Man (if one continues in this spirit) would be the ideologue or apologist of power *par excellence.* For de Man these stories of who did what to whom, and of an earth that can be mapped according to a tale of distribution, resources, acquisition and ultimately of justice and injustice are just that—stories—that allow us to make sense of inequality. Let us say that for de Man narratives of *inequality* or disproportionate distributions are ways of spatializing and temporalizing (or englobing) what cannot be narrated because there is no common space or measure. True, as long as we remain at the scale of world history then one human's encounter with another takes place within historical time (a time of nations, revolutions and corporations), and within global time: the time of this humanized earth that is lived as an environment in which natural time (a time of ecological bonding) is now being interrupted by capital time, a time of resource seizure and depletion that will end the world for us.

Does this just require us to add another political scale, to add the dimension of the Anthropocene or earth system science to the list of theoretical contexts? We could say that *Heart of Darkness* is prescient of a certain expansion of the political to the environmental and the planetary, for what Marlowe encounters and ideologically witnesses as a horror that must somehow be saved by a narration of the higher civilizing good of empire is ecological rapacity, and the limits of the global imaginary. The Africans, the ivory, the Congo itself are all so much standing reserve for a journey of acquisition that is both covered over, but ultimately recognized by, the synecdoche of European learning that is Kurtz. The colored map, the moral projects, and the dreams of order are all humanizing masks for ecological destruction. Ultimately the culprit is empire, and ultimately what suffers and feels the wound is all the earth, which appears pale and *at an end* if one confronts the lie of European goodness too directly.

What if there were another shift of scale, not towards the global, not towards accounting for self and other by way of the narrative of the human earth (where existential conflicts are seen as fragments of a larger narrative of capital and acquisition), but where this narrative of the human earth—of capital, conquest and then the possibility of redemption by way of an outside of capital—were a way of symbolizing what is outside narrative and outside of a certain dialogic conception of justice? This is why I think de Man might be one of the few theorists capable of approaching what I would refer to as the hiatus of catastrophe.

For the most part, history (including human history) has been one of utter contingency, violence and volatility, and thoroughly inhuman: the brief era of the publicly distributed book and the private sense-making-imaginary of deep attention coincides with the era of high industrialization, which is also the era that will precipitate the Anthropocene as a readable mark on the planet. That is, there is a brief period when we might believe in a nature that is lawful, benevolent, stable, not subject to change, and—as Kant suggested in reflective judgment—capable of being viewed *as if* it were in accord with a just and virtuous narrative. We might also think of history as readable, as having a face, direction, intentionality and sense, but only if we have tamed and stabilized the earth for a form of techno-science that relied on the theft of energy from another world that was not within the enlightenment or human purview. Only exceptional industrialization of agriculture could yield such a benevolent, cyclic and environmental nature. By exceptional I mean short-lived, and available to some humans only. This exceptional suspension of catastrophe generated something like the political scale as literary history has known it: we read a text within a history of capital, and that capital is semantic, traced back to persons and practices, and organized sensibilities.

When Lacan talks about the Imaginary as the way individuals live the Symbolic's relation to the Real, such that we imagine that we are thoroughly capable of grasping the full complexity and contingency of existence with our quaint anthropomorphic figures, we can say that there is no relation or correlation, no extension or continuity between the linear and human sense we make of what we take to be our world, and the multiple, volatile, and infinite forces that operate with blind disregard for human sense and intentionality; there *are* narratives about that disjunction, a narrative that will present an opposition between the dark chaos of contingency and the homely order of law, but there can be—as de Man says - no theory of this narrative (de Man 1971, 271-72). It is that disjunction that we resist in all forms of theorization that would give an account of the temporal or spatial relation between that which resists all narrative order and some way of accounting for this order by way of notions of history, colonialism, capitalism and so on. Now one might want to set up a narrative about this type of theory, and it might go something like this: "theory" began with an initially radical emphasis on language and textuality, but became far too confined to formalism and lost politics. With the turn away from deconstruction to theorists like Gilles Deleuze and Félix Guattari more attention could be paid to life, affect, bodies *and politics.*

When the Anthropocene was proposed, this post-textual direction of theory (towards life or towards a new materialism) seemed to be confirmed and intensified. It is because the planet (for us) has been destroyed that there must be a we, and this "we" would be a new humanity united by emerging from what might otherwise appear to be defeat. It is supposedly materialism, the brute reality of ecological destruction and the imminence of our demise, that will save us, generating a humanity to come. The recuperation of the Anthropocene by way of the sublime has allowed the trace of human blindness and a radically impolitic event to reinvigorate politics. If, however, the Anthropocene were *not* recuperated by this mode of sublimity—this intimation of a greater humanity that emerges precisely in the moment of its vanquishing—one might begin to *read.* In this case, geological inscription and stratification would not be coopted by a narrative of vitality, and certainly not of a *living system.* The image of the earth as a living system is a restriction of scale to the narrative that accounts for life, but this same earth might also be viewed from no scale at all, without the imminence of life operating as a frame. Before turning back to de Man and his claim for a mode of sublime that deprives matter of all face, life and homeliness, we might note that something of this conception of a pre-vital and pre-physical matter was articulated by the same Deleuze and Guattari who have been hailed as the great masters of vitalism. Speaking through the character of Professor Challenger (after Lovecraft) they write:

> He used the term *matter* for the plane of consistency or Body without Organs, in other words, the unformed, unorganized, nonstratified, or destratified body and all its flows: subatomic

> and submolecular particles, pure intensities, prevital and prephysical free singularities. He used the term *content* for formed matters, which would now have to be considered from two points of view: substance, insofar as these matters are "chosen," and form, insofar as they are chosen in a certain order *{substance and form of content).* He used the term *expression* for functional structures, which would also have to be considered from two points of view: the organization of their own specific form, and substances insofar as they form compounds *(form and content of expression).* A stratum always has a dimension of the expressible or of expression serving as the basis for a relative invariance; for example, nucleic sequences are inseparable from a relatively invariant expression by means of which they determine the compounds, organs, and functions of the organism (Deleuze and Guattari 1987, 43).

It is hardly remarkable to talk about a pre-organic matter, nor a pre-vital matter but it is worth taking note of the notion of matter as pre-physical, as that which would be prior to the opening out of space and time. Such a materialism would be pre-political, and if the emergence of the polity were to be narrated it would be able to do so only by manufacturing a certain plane, which would necessarily leave out the radically material dimension prior to physicality. It is this sense of matter—inhuman, purely intense, devoid of homely sense and affect—that de Man describes when one regards matter (after Kant) "as the poets perceive it" (1996, 81). By contrast, to see into the life of things, or to intuit the vitality and vibrancy of matter is to give it a face or form; if such a vision is political (by locating what is seen into a narrative of human relations), it achieves this politics by way of privation or defacement.

I think this is what Deleuze and Guattari referred to as minor literature: an inscription or matter that is so minimal, so cramped, so disconnected from all context, world and sense that is it neither human nor animal (though possibly capable of becoming-animal, indicating a potential outside constituted organisms). Here is de Man, first on language as a mode of silencing or rendering mute, and then on animation as a mode of defacement:

> To the extent that language is figure (or metaphor, or prosopopeia) it is indeed not the thing itself but the representation, the picture of the thing and, as such, it is silent, mute as pictures are mute. Language, as trope, is always privative. Wordsworth says of evil language, which is in fact all language including his own language of restoration, that it works "unremittingly and *noiselessly.*" ... As soon as we understand the rhetorical function of propopopeia as positing voice or face by means of language, we also understand that what we are deprived of is not life but the shape and sense of a world accessible only in the privative way of understanding (de Man 1979b, 930).
>
> ...the challenge to understanding that always again demands to be read. And to read is to understand, to question, to know, to forget, to erase, to deface, to repeat—that is to say, the endless prosopopoeia by which the dead are made to have a face and a voice which tells the allegory of their demise and allows us to apostrophize them in our turn. No degree of knowledge can ever stop this madness, for it is the madness of words. What *would* be naïve is to believe that this strategy, which is not our strategy as subjects, since we are its product rather than its agent, can be a source of value and has to be celebrated or denounced accordingly (de Man 1984, 122).

This leaves us with what might appear to be the depoliticizing abandonment *par excellence*: when you say empire, colonialism, patriarchy, and the blow corporate techno-science bears to the planet, I declare your reckoning of accounts to be human all too human, and then demand that we face up to the radical inhumanity of that which we tend to grasp (narcissistically) as human. This is perhaps why the critical impact of notions of anthropogenic climate change and the Anthropocene has been enlivening, rather than devastating; precisely when we ought to be confronted with "civilization" as a trajectory of wreckage, we become all too focused on surviving. Far from recognizing the ways in which desires, intentions and an epoch of humanism, enlightenment and globalism have destroyed their own conditions of emergence, the overwhelming response has been an insistence on hope for the future (whether by way of politics or geo-engineering).

In response to this recuperation, I want to conclude by focusing on a rigorous sense of aesthetic ideology, with an emphasis on *aisthesis*, or the transition from what is given—the sensible—to what that givenness presupposes or demands that "we" assume. To insist on *the* political, at any level, is to posit a putatively legitimate register that would be the appropriate milieu for taking an account of any event. Where the political tended, once, to be determined by way of socio-economic relations, it is perhaps—today—shifting to a different register of affect and corporeality, or how bodies are formed with the desires, interests and practices that *then* allow for individuals to form social wholes. But what if those practices of political theory were themselves dependent upon an epoch of suspension, in which the earth, the globe, nature, affect or corporeality—or even humanity—could appear as an object of stable knowledge only with certain practices and formations that would precipitate the destruction of the milieu on which they depend? There can only be the polity with certain forms of life, and from the very outset stable political wholes relied on violent forms of inscription and appropriation that erased and defaced in order to institute a plane of history. What if the (reflective and critical) reading and theorizing we direct to ourselves were the outcome of an era of technologies of the eye, hand and an industrialized relation to nature, with these technologies in turn always borrowing from the earth's reserves in an ongoing debt that can never be discharged? What if what we know as politics—the practice of tracing what appears as contingent, universal or natural back to human forces—were possible only in a brief era of the taming of human history? There could only be "the polity" with the sense of nature *not* as a force in its own right, but as an environment, as nothing more than the passive background for praxis. Today's climate change is an acute event in a long history of changing climates; what we take to be "our" earth is a brief and unsustainable period of human-friendly ecosystems that have been manufactured violently and blindly, always at the expense of much life deemed to be non-human.

What if today's threatening conditions of volatility and disaster were to make anthropomorphism and the belief in nature increasingly impossible? What sort of politics would we be left with in this disfigured world without face?

I want to conclude, then, by thinking of politicizing what we have taken to be politics, or—perhaps—thinking about the epoch of the political. Before returning to de Man I would suggest that there is precedence for this in Theodor Adorno and Max Horkheimer's *Dialectic of Enlightenment* (2002), where the period from Homer to the present is unified and depicted as an epoch of the critical subject, for whom the chaos and terror of the world is increasingly stabilized by technocracy, and for whom exercises in demystifying formal abstraction have a certain pacifying mythic (and delusional) quality. Similarly, though with key differences, Bernard Stiegler (2008) also regards the critical subject of reading and privacy as inextricably tied to textual, mnemonic technologies and, like Adorno and Horkheimer, sees that same critical subject as intertwined with (and blind to) the inscriptive and figurative systems that reduce the complexity of what comes to be known as nature to a manageable and manipulable substance.

In conclusion, then, I would align de Man with a counter-political politics: any passage or transition to *the political* as the legitimate context for making sense of inscriptive systems must occlude the forces that generate the critical, politically astute subject-reader. Today we know those forces to be part of a destructive network of technologies that generate power (and power in a quite literal sense, including the depleting power of finite fuels, and the depleting power of the critical archive as individuals take on new modes of relating to texts and images). What de Man allows us to think about in this bracketing of the epoch of critical man is precisely that aspect of his work that has always appeared as so scandalously counter-political: if there is *nothing* to legitimate the transition from inscription to sense, then politics is not so much about expansion—explain the text by way of its being an epiphenomenon of a grander or global whole—but about collapse.

How might such an event of collapse help us with politics today? It would refuse the prima facie goodness of such a question: there would be no self-evidently justifiable "we," no unquestioned right to sustain the polity, or the political. It would then be possible to look to the technologies—primarily inscriptive and aesthetic—that generate the beautiful soul of critical reading, and the equally beautiful ecology and polity that is its natural other. What we then might have is a genealogy and geology of the event of aesthetic ideology. The reality that de Man scandalously claimed would remain unreadable, or readable only by way of allegorical delusion—the natural world from which the text emerges and which could always be retrieved and re-read—would be the outcome of an epoch of affluence and climate stabilization that may now be coming to a close. If it is not coming to a close—if, say, we are about to continue this hiatus of disaster (for some by way of geo-engineering)—then perhaps it would be more imperative than ever to think about the generation (in all senses of the word) of this human-nature. That is, we would need to think about the generation of power (all the resources of human and non-human energy that enabled the emergence of "humanity"), the generation that "we" are, in terms of the debt and destruction bequeathed to what might remain of a future, and the generation or genesis of the stratifications that allowed a figure of "nature" as our benevolent whole. This would also include questioning the generation of humans who are beginning to lament the loss of deep reading.

4. The Geological Sublime

Bruno Latour has argued that the problem of scale of the present crisis —which he defines as the ecological destruction that will be labeled as the Anthropocene—requires a reversal of the sublime: if the Romantic sublime followed from the magnitude of nature and its indication of an everlasting and eternal power, it is now the case that nature is puny, fragile, destroyed and limited compared to the colossal force of man: "We realize that the sublime has evaporated as soon as we are no longer taken as those puny humans overpowered by 'nature' but, on the contrary, as a collective giant that, in terms of terawatts, has scaled up so much that it has become the main geological force shaping the Earth" (Latour 2011, 3). For Latour, it is the materiality of the crisis –that nature appears as all too finite, limited and capable of a temporal end—that should preclude us from referring to the present crisis as sublime.

Latour is interested in exactly what concerns me in raising the importance of the sublime: the destruction that presents itself to us cannot be presented. For Latour, though, this is an institutional problem, and one of discipline. The very disciplines that allow the earth to be viewed and studied, by virtue of being disciplines, limit the range and scale of what is studied: "There is no single institution able to cover, oversee, dominate, manage, handle, or simply trace ecological issues of large shape and scope. We have problems, but we don't have the publics that go with them" (Latour 2011, 1). The sublime concept of Nature—this ineffable "x" that lies as the ground from which the viewed world of relations and things emerge—must be abandoned. Abandoning that infinite and ineffable nature is the first stage of what Latour puts forward as a project of composition. We are waiting for Gaia: not nature as some absent sacred beyond, but nature as a composed, interconnected, and dynamic unity that is constituted from a series of modes of existence, including multiple human practices. We must start to think of ecology *without nature.*

Why? For Timothy Morton (2007) the notion of nature as eternal and everlasting—a nature that would provide some ground for our nostalgia and yearning—never existed. What we thought of as nature was created or effected (ex post facto) from a series of forces and interactions, some of them human and some of them not. What Latour referred to as the modernist split between the world as it is and the world as it is known, needs to be displaced by one level of interacting forces—humans and non-humans, delimited things and a series of less thing-like entities (including odors, atoms, waves, particles and habits). For Latour, scientifically, it is better to think of the world neither as a matter that is represented, nor as the effect of construction, but as a composition that follows on from "matters of concern." Each discipline, practice or mode of existence is affected by the world (and this power to be affected follows from the fact that the world *matters* or possesses a degree of force), but the response to that affect is given differently in the range of habits and practices from which we too are composed.

It follows then that there is no "we" as such, and that a compositionist response would require a coming into relation of various modes of existence. To think of nature as sublime, as an immensity that can be indicated but never presented is not only defeatest—allowing perhaps for the fetishization of a pristine nature that never was and never can be—but also undisciplined: the only way we have a world at all is through the concerns it prompts us to act upon, and it is that prompting that enables discipline in two sense. Disciplines are active, knowledge-composing, shared and adaptable practices that allow a common world to come into being, creating modes of existence that distribute distinct powers to be affected. Discipline is also what is required to confront the problem of scale; if we don't compose some interdisciplinary Gaia then we will not have the power to be affected that Latour places at the heart of practice. Discipline (in the sense of knowledge practice) is at the heart of composing the world; and it would be from that composition that one might also then garner the discipline required to act upon the matters that would now concern us. What Latour refers to as sublimity is both no longer possible—for it is evident that nature does not overpower us and offer itself as eternal and everlasting—and certainly not efficacious. At present the panorama of possible wreckage that confronts us *may* seem to be destructive of any possible public, but rather than express resignation and attribute sublimity to the spectacle that intimates a scale beyond our ken, we need to compose a unity that connects rather than disconnects our affects and actions.

Taking Latour's and Morton's arguments into account, and recognizing the value of composing (but not constructing *ex nihilo*) some interconnected whole that refuses any noumenal presence that would lie behind and explain the world of relations that we witness, what I am about to propose would present at least three problems. I am going to suggest that we pick up Paul de Man's concept of the material sublime, and that we do not isolate the ecological crisis from a series of other crises, including the financial crisis, the war on terror, the new modes of terror (including bioterrorism) and the specter of mass viral pandemic. I would not be alone in gathering these catastrophic risks together; but apart from philosophical interest I would suggest that doing so has practical import. And it is the problem of practice that is tied to the sublime. On the one hand one must act *as if* there were a distinction between the material world of causes, and the decision of human freedom: praxis must at once be oriented to the world but not of the world. It is that idea of freedom that remains in Derrida's "democracy to come," and sustains Latour's emphasis on the world being both composed from our concern, but also constrained by the matters that prompt concern. If the world were fully knowable then there would be no room for human freedom; I certainly cannot theorize or give an account of freedom, for any description of freedom would necessarily require some narrative account, and would thereby fall back into causal logics. However, if we were to accept the force of the sublime—that there is indeed some "x" that is unrepresentable, untheorizable, and beyond any possible experience—then there would be something like a freedom effect. One might think of this in two ways. Derrida, insisting on the future *insofar as it is futural,* asks that that we open ourselves to the "to come." Freedom would not be a theoretical object, but it would have practical effects: act *as if* the future were incalculable. Indeed, responsibility would only be responsibility if it were responsive without having calculated its range in advance.

Alternatively, de Man's "material sublime" would shift sublimity from the quickening of the subject's powers, and would seem to de-activate or paralyze thinking. It would be a sublime without Idea. It would also be a sublime without Latour's to-be-composed public, and without Derrida's other. Such a sublime would be aesthetic in de Man's sense *not* because it has to do with art and composition, but because it would propose a mode of seeing *without sense or teleology.*

Why, outside of literary theory, would one want to exacerbate Latour's problem of the absence of a public? Surely we want enabling notions, and ways of making our world manageable? Only a restriction into a narrow disciplinary frame of high theory (and de Man's high theory at that) would warrant such a strategy, and it is precisely *that* disciplinary myopia that—it might seem –has done nothing to help the practical task of facing twenty-first-century crises. But I would argue quite the contrary: what is required is neither the connectedness of composition, nor the hyperethical investment in the absolute singularity of every *person*'s world. It would only be with *impersonality*—when it is not the face, affective force or empathetic life of the other—that sublimity might be approached, and the approached may no longer be moral or ethical but pragmatic. What if we could look at all forces with the eye that is not detached *from "the world"* but is confronted with decomposition, fragmentation and detachment *tout court.* There would be no other, no humanity, no "us" deemed immediately worthy of survival, but the question would then finally be posed: what calls to be saved? Is saving, surviving, and living on a prima facie value?

If the Anthropocene offers itself as too big a problem for the isolated disciplines and institutions we have, what on earth would we gain from throwing it in with anything and everything that might bring "us" to an end? I would suggest that thinking about the sublime, and thinking about the material sublime in de Man's sense, requires and enables us to create a new conglomerate, a totality that is not a dynamic, creative and open whole of overlapping but distinct disciplines, but rather a force of destruction or dis-unification that is single only in its lack of quality or distinction. For all that has been said about deconstruction, difference and politics over the years—with deconstructive difference having been aligned with race, sex and other identity differentials—I would argue that de Man's material sublime offers a more productive (because destructive) theory of *indifference,* and this is so even if it is de Man who seems to be the most guilty of reducing difference to textual difference.

The Romantic sublime occurs when my ability to comprehend or represent is challenged by what appears to be an overwhelming or infinite nature. The self of sensation and material existence is vanquished, generating the feeling one has of oneself as immaterial. The relation between sublimity and ethics is human but impersonal: the thought of acting without regard for actual humanity, and being motivated only by pure duty *then* allows for the thought of elevation. Then and only then does one perceive a human action from duty as sublime. In Kant's example: any notion of giving up one's life for an end within this world, ranging from suicide to mere heroism, merely subjects my individual existence to a finite end; however, if I choose to die for a duty that exceeds all worldly ends, then I disclose to myself the degree to which I am a supersensible member of the kingdom of ends. Such a sublimity is Romantic because it operates with at once a striving for infinite apprehension, and then a feeling that the very inability to experience the infinite generates a sense of that which lies beyond the given and finite. In Wordsworthian terms: this nature here, generates a sense of something far more deeply interfused, a 'still, sad music of humanity' that is quite different from the actual presented world of humans in existence. There is an intimation of the infinite that is generated from the feeling of the finitude of any sensible or material life.

De Man notes the *apparent* similarity between Kant and Wordsworth but then spells out an important distinction to do with a contrary relation to affect: whereas the Wordsworthian sublime is humanizing, with nature appearing as if harmoniously in accord with our higher nature, Kant's version is architectonic. There is something depersonalizing and dehumanizing in the Kantian mode. To view nature as a vault, or as if built, is *de-naturing or de-naturalizing,* if by "nature" we think of some living, feeling, composed and affecting unity. To view nature architectonically is, I would suggest—and this is de Man's line of argument—to view the world as if humans were not present, *and* as if the world were not a living being so much as assembled, built and *unreadable.* De Man quotes Kant, and compares him to Derrida's conception of the architectonic nature of the sublime. For both Derrida and Kant, de Man notes, one needs to think of sublimity as *erection,* and this yields a specific mode of time and our relation to time; this is not a time of our own mastery, for the very symbol of virility—the pure erection without the "languid, yielding" affects of the beauty—is "anything but what one—or should I say men?—think(s) it to be":

> What is it for Kant? We receive a hint in a passage which tells us how to look at the sublime, how to read judiciously, like the poets ("*wie die Dichter es tun*"): "If we call sublime the sight of a star-studded sky, we must not base this judgment on a notion of the stars as worlds inhabited by rational beings, in which the luminous points are their suns, moving purposefully and for their benefit. We must instead consider the sky as we see it [*wie man ihn sieht*], as a wide vault that contains everything. This is the only way to conceive of the sublime as the source of pure aesthetic judgment"...
>
> Kant's architectonic vision here appears in its purest form. But a misguided imagination, distorted by a conception of romantic imagery, runs the risk of setting the passage awry. It may appear to be about nature in its most all-encompassing magnitude but, in fact, it does not see nature as nature at all, but as a construction, as a house (de Man 1996, 126).

Already, then, within the Romantic sublime there are two tendencies that de Man outlines by way of the relation between the aesthetic and the political (and there is also a sexual register whereby the erectile image of self-inflation is ultimately a submission to a volatile force not of one's own governance, and liable to dysfunction). The aesthetic is inscriptive, textual *and sublime because the materiality it presents has not been humanized,* nor rendered living, nor imagined as a wholeness or a connectedness. Commenting on Hegel, de Man writes:

> Monuments and statues made of stone and metal are only pre-aesthetic. They are sensory appearances, all right, but not, or not yet, appearances of *the idea.* The idea appears only as written inscription. Only the written word can be sublime, to the precise extent that the written word is neither representational, like a perception, nor imaginative, like a phantasm (de Man 1996, 110).

In his essays on Kant's and Hegel's conception of the sublime de Man works tirelessly to orient the sublime towards a vanquishing of the imagination, or to arrive at something like an *imagination without image* that would—for that very reason—have to be a mode of disaffected calculus: "The faculty of imagination is itself beyond images" (122). This means that the sublime would be an appearance of a certain non-appearing, a viewing of something not as an image that stands in for and connects what is viewed with life and spirit, but "reestablishes contact with the classical philosopheme that Kant inherits from Leibniz, in which the homogeneity between space and number, between geometry and calculus, is to be reestablished by ways of infinitesimal motion" (122). One might make a connection here to the ways in which Deleuze and Guattari draw on the distinction in art between abstraction and empathy: the former mode is—as de Man describes the Kantian sublime—one in which the world or what is presented is given as pure form, quite distinct from life (pyramids, straight lines, patterns); and this is opposed to empathy where one feels and perceives the life of what is presented. To quote Worsdworth: "we see into the life of things." The difference between *monuments* and inscription is that monuments recall or stand for something, and to this extent one attends to what they present, whereas the word or inscription *as word* institutes a separation, or *de-natures* us; we no longer perceive the world as ours, as representative. Mind must not be welded seamlessly to some enchanted world that is perceived as ours, as living or as meaningful: rather, it is in *not* being the world, in being detached, separated—as if the world were mindless and inhuman—that sublimity emerges: we no longer live in a world populated by spirits.

Rather than passing seamlessly from what is presented to the proliferation of what it signifies, we perceive the presented and think critically about it *not* being that which it signifies, such that it is the distance and distinction of the word that renders it sublime. We would not perceive the oceans as a symbol for flourishing life, but just as though they were *put together* (not intelligently designed so much as placed). Is this not how we might imagine the human world and archive (after humans) in all its forms: not as something that we would read, or that would be readable? The geological sublime is therefore the challenge of looking at the entire archive of the earth—including human script—as one might look at the marks left on buildings by the forces of weathering. How would we read ourselves if we were not to assume some ultimate readability or spirit beneath the materiality of text?

If one looks at the post-apocalyptic sublime the reverse takes place: we imagine the detritus of the world after humans as ultimately all that is left of a world that is nothing more than readability: it is as though monuments, or the signs we give to ourselves as beings of freedom, constitute humanity and the only capacity humanity has to think beyond itself. We can think of two sublime-tragic instances: at the conclusion of Danny Boyle's *28 Weeks Later* (2007), after a seeming containment of the "rage" virus "we" witness hordes of infected post-human humans rushing upon the Eiffel Tower: "we" see the human world as if it were no longer populated by beings who would be able to recognize it as human. We see "our" world of freedom, sense and monument from the point of view of beings who can only look at the world as no *world* at all. To lose that readability would be to lose the world. And yet it is precisely that which (de Man's) Kant seeks to find in the sublime, a gaze as if this world meant nothing, as if we could view nature not as monument but as vault. Consider the sublime scene from the original *The Planet of the Apes* (1968) where Charlton Heston finds a fragment of *The Statue of Liberty* left on the beach *as if it were so much random junk*. He now lives in a world in which the very symbol of self-erecting freedom is no monument at all, merely contingent waste washed ashore. For de Man, Kant suggests that the look we might direct to the world *as* just something assembled, is sublime. Such a look would detach us from our world of feelings and meanings. But of course *The Plant of the Apes* does not view *The Statue of Liberty* in that manner; indeed, it is characteristic of the post-apocalyptic sublime to regard the loss of readability and (human) meaning as the loss of the world as such.

Chapter Three

Reading Paul de Man While Falling into Cyberspace In the Twilight of the Anthropocene Idols

J. Hillis Miller

In Memory of Cat Kiki

1. The Linguistics of Literariness and Ideology

> *What we call ideology is precisely the confusion of linguistic with natural reality, of reference with phenomenalism. It follows that, more than any other mode of inquiry, including economics, the linguistics of literariness is a powerful and indispensable tool in the unmasking of ideological aberrations, as well as a determining factor in accounting for their occurrence. Those who reproach literary theory for being oblivious to social and historical (that is to say ideological) reality are merely stating their fear at having their own ideological mystifications exposed by the tool they are trying to discredit. They are, in short, very poor readers of Marx's German Ideology.*
>
> Paul de Man – "The Resistance to Theory"

This chapter is a long commentary or endnote to the passage from Paul de Man's "The Resistance to Theory" cited above as my epigraph. The reader will see why and how as the chapter progresses.

My questions in this essay are the following: In these bad days, what good is studying literature or literary theory? What good can be served in our present dire situation by appropriating Paul de Man's procedures of "rhetorical reading" to help understand our present condition? By "these bad days" (a citation from Matthew Arnold's "To a Friend" [Arnold 1965, 105]), I mean the many ways things are not going at all well in the world, and in particular in the United States. Studying literature and literary theory would be useful only if some form of language, including digital sign systems and visual images, is in one way or another involved in the ways we live now. De Man argues cogently that we must learn to "read" visual images, and I'll return in the third section of this essay to that somewhat counter-intuitive claim. Tom Cohen's brilliant essay for this volume has forcefully demonstrated that climate change and the other features of our life that I shall now identify are linguistic events.

1) We are experiencing global climate change that may soon make the species *Homo sapiens* extinct, after putting our coasts and coastal cities under water (New York City, for example, not to speak of Florida and my coastal home on Deer Isle, Maine). California is already becoming a desert or dust bowl. The Mid-West has had frequent violent storms with tornadoes, alternating with droughts and heat waves. The winter of 2014-15 was the worst on record, and violent storms are sweeping the Mid-West as I revise this in late April, 2015.

Nevertheless, many people still fiercely deny "anthropogenic" climate change. (See Cohen's essay for the deceptions buried in that word "anthropogenic.") The facts about this are amazing, to me at least. They are strong evidence of the power of ideological lies to influence belief and behavior. An essay in *HuffPost Politics* for April 23, 2015, for example, reports that

> A Gallup poll released Wednesday [April 22, 2015] shows just how resistant some Republicans are to the science of climate change. In polling conducted over the past five years, 59 percent of self-identified conservative Republicans said they don't believe that climate change is happening now, and 70 percent said they don't believe humans are responsible for it. Gallup asked about 6,000 Americans of diverse political ideologies whether the effects of global warming would be felt in their lifetimes, in future generations or not at all. Forty

> percent of conservative Republicans deny that global warming will *ever* happen, while an additional 19 percent believe that it will only affect future generations. However, a May 2014 governmental report found climate change is affecting all areas of the United States. Although there's a wide scientific consensus that climate change is caused by increased levels of carbon dioxide in the atmosphere, most conservative Republicans reject the idea that there is a link between pollution and rising temperatures. They are the only political group to have a majority not believe in the connection. (Sola 2015).

When I worked on this essay in an earlier form in July and August, 2012, the United States was experiencing the hottest year on record. Fires had raged in Colorado and elsewhere. A severe drought was destroying our corn and soybean crops. Food prices were already going up. Immense *direchos*, or thunderstorm fronts, had devastated parts of the Midwest and East, spawning tornadoes, downing trees, causing floods, and widespread power outages. A sudden melting of the entire Greenland ice-cap surface was reported in 2012, to the amazement of scientists, who keep saying: "Oh dear! All this is happening much faster than we thought!" Scientists also had just discovered, to their further amazement, that the huge thunderstorms are pushing warm air up into the stratosphere and posing a threat to the ozone layer over the United States (Fountain, 2012). New reports or events of this nature have occurred almost daily since then. Most of my new evidence in this introductory section comes from just a few days of reports in the digital media in late April of 2015.

This long series of climate catastrophes has in fact accelerated in frequency and intensity since the summer of 2012, for example with Hurricane Sandy in late October 2012, which flooded large sections of New Jersey and Manhattan. Hurricane Katrina, that devastated the gulf coast, had already occurred in 2005. Hurricane Sandy confirmed that the prediction that climate change could cause flooding of the United States East Coast was not a science fiction fantasy or a wacky prediction of hysterical climate scientists.

The Antarctic ice is rapidly melting irreversibly. This will ultimately raise sea-levels worldwide about fifteen feet, as numerous recent on-site investigations have discovered.

Another (counter-intuitive) example is the recent extremely cold and snowy winters in the United States, the worst on record in many places all the way down to the Gulf Coast, where snow and ice-storms of an intensity such as have occurred are exceedingly unusual. These winters have been accompanied by so-called "polar vortexes," a beautiful figure of speech. These ferocious winters would seem to support the claims of climate-change deniers. ("You see, the weather is just as cold or colder than ever! Who says the climate is getting warmer?") Scientists, however, cogently relate these cold winters to the melting of Arctic ice caused by global warming that is registered in thermometer records (another form of language). An essay in *Scientific American* for May 2015 by Mark Harris persuasively and succinctly states the evidence for this connection:

> Accurate Arctic climate models are critical. [For most human beings, changes in the Arctic exist as "linguistic" "models." Few of us have actually seen the open water in the Arctic. Television videos are after all another digital artifact, therefore mediated by a form of language.] Scientists at Pacific Northwest National Laboratory think that less sea ice in the Arctic means heat energy that is usually trapped underneath this ice can escape into the atmosphere. This rising heat can disrupt the jet stream, the swiftly moving high-altitude air current that makes it quicker to fly from west to east in the U.S. than the other direction. Some scientists believe that the jet stream acts as a barrier that prevents frigid Arctic air from moving south and that changes to it can cause extreme "polar vortex" weather events such as those that have frozen East Coast [and Mid-West I add. jhm] cities in the U.S. during the past two winters (Harris 2015, 66).

Meanwhile such self-destructive activities as widespread use of tar sands and increased use of fracking for shale gas continue at an ever-accelerating rate. Drilling for new oil wells in Oklahoma has made that state the scene of constant earthquakes caused by the injection of water (produced by the oil well drilling and by fracking) deep into the earth. An essay by Clifford Krauss in *The New York Times* reports in a jubilatory mode that the United States is now the world's largest producer of oil and gas. "Shale drilling," says the lead, "has not only transformed the United States from a dependent consumer into a robust producer of oil, it is also transforming the price dynamics of the global market" (Krauss, 2015). Well, three cheers for that! Never mind that this is just contributing further to planetary "ecocide" (Cohen's word).

We have, let's "face" it, in any case already passed the tipping point of irreversible climate change. Language is deeply involved in this happening. It has been facilitated in part by climate change deniers, who believe the lies told by politicians and the media claiming that the evidence for "anthropogenic" climate change is a hoax perpetrated by "mad scientists," against the evidence. Another example is the ads by the Petroleum Institute of America still (April 2015) shown nightly on NBC Evening News. These ads are supported by appropriate graphics: cross sections of a schematic fracking site looking clean and neat, crowds of happy workers, a superhighway full of speeding cars and trucks. In the foreground an attractive young woman claims that fracking is entirely safe. It will create millions of new jobs, and make the United States happily "energy independent."

If you have the strength and have never yet done so, just try watching one example of the NBC Evening News from 6:30 to 7:00. As a testimony to the state of our culture, it is a dismaying mixture of 1) violent segments of so-called news calculated to keep viewers in a constant state of anxiety and terror: police murders of innocent unarmed "suspects," volcanic eruptions, replays of the Boston Marathon Massacre or of the attempted assassination if President Reagan, or of the latest violence in the Middle East such as another town destroyed in Syria, Yemen, or Afghanistan. The same "shots" of violence are shown over and over; 2) a seemingly endless sequence of brief ads telemarketing patent nostrums for things like erectile dysfunction, or painful intercourse during menopause, or insomnia, or arthritis pain, or Irritable Bowel Syndrome, or diabetes, or Alzheimer's, all mixing videos of happy people taking the medicine in question with screen inscriptions and voiceovers touting the virtues of each, plus the intoning of side effects, as required by law (internal bleeding, heartburn, heart attacks, stroke, diarrhea, muscle pain and weakness etc.): "If you have an erection lasting more than four hours see your doctor right away." "Ask your doctor if Crestor [a particularly dangerous statin] is right for you." Anyone who listens at all carefully to the side effects would be either truly desperate or an idiot to buy any of these drugs. A tremendous amount of poetic inventiveness goes into those ads, as in the one that shows a little dog searching for a safe place to hide a bone. It is an ad for an investment firm.

A de Manian rhetorical reading of the segments in their sequence of a given half hour of the NBC Evening News would demonstrate the basic claim of this essay: that de Manian procedures would help us at least to understand the plight we are in as the victim of lies and baseless figurative transfers. The ad for fracking takes its place in the series on a given night. Actually two different versions exist and are often both shown in the course of a given half hour.

A lie, for example those in the ads I have just mentioned, or the claim made by many Republican politicians that climate change is a hoax, is a peculiar form of speech act. A lie's referential value is nil, but if it is believed in and acted on it is an efficacious or "felicitous" speech act. Much of the linguistic involvement in our present unhappy situation is made up of such lying speech acts.

2) The world is still recovering from a deep global recession. This was caused in large part by the subprime mortgage crisis, that is, by criminally fraudulent actions by our biggest banks and investment houses that would have been impossible without language in various forms. An example is the Ponzi Scheme computer programs creating by a regressive series ultimately worthless "credit default swaps." The banks are now again just doing fine and are up to their old tricks again.

Though employment is way up, we still have relatively high unemployment in the United States, especially if those underemployed, part time, or self-withdrawn from job-seeking are included. Many of the new jobs are low-paying service positions, exacerbating the already huge discrepancy between what the top 1% make and what an ordinary worker in, say, WalMart, makes per hour. In many European countries unemployment is even higher, and several countries (Greece, Spain, Ireland, Great Britain) have slipped back into deep recessions or into economic stagnation. Austerity measures, just the wrong strategy, have been adopted in many countries, such as the four I have mentioned, with disastrous results. The argument for austerity rests in part on a false figurative transfer equating family finances with nation finances. Families should not spend beyond their means, so neither should countries. The two are not the same, partly because countries print their own money, which families do not. Nations can, within reason, safely run up large deficits.

In the United States the disparity between the income of the top one percent and the remaining 99% has never been higher. The richer are getting richer, the rest of us poorer, partly as a result of self-destructive changes in our tax laws under President George W. Bush and other previous presidents. The United States is becoming more and more like a third-world country in its wealth distribution, with a concomitant gradual disappearance of the once prosperous middle class.

The global financial meltdown was brought about by the folly and greed of our politicians and financiers. Little has yet been done to re-regulate the banks and other financial institutions, for example by prohibiting them from speculating wildly with investors' deposits or by charging them a tiny fee for each transaction to discourage microsecond automatic computerized trading (another quasi-linguistic procedure). The Libor scandal of several years ago now over the rigging of the interbank lending rate is only one example of bank officials' criminal fraud. That fraud has so far gone for the most part unpunished, though a number of settlements in the billions with fraudulent big banks have at lest given them a vigorous slap on the wrist. Paying a good many billions in penalties is a piece of cake for them, however, part of the cost of doing business.

Many experts confidently predict that eventually another financial meltdown worse than the first will occur, if nothing is done soon to regulate the banks and other financial institutions. This new regulation seems unlikely, since our present governments are in cahoots with the banks and still see them as "too big to fail."

3) In the United States rightwing lobbyists and media propagandists to a large extent control Congress and public opinion, particularly since the Republicans took control of both houses of Congress in the 2014 elections. Republicans are committed to stopping anything President Obama tries to do. They are funded by seemingly limitless contributions from super-wealthy extreme right-wing Americans like Sheldon Adelson or the

Koch Brothers, and by corporations. The latter are now absurdly labeled by the Supreme Court as "persons" with the same constitutional rights to secrecy about their political contributions that are guaranteed to persons. We have highly influential news media in the United States, such as *Fox News* and many radio talk shows, that are more or less lying propaganda arms of our right-wing parties. Nevertheless, they are believed by many citizens to be telling the truth. An example is the attacks over the years of the Affordable Healthcare Act, which has been a great success, increasing by around fifteen million American citizens those now getting affordable healthcare. It would be even more if so many Republican states had not perversely refused the expansion of Medicaid written into "Obamacare," thereby condemning some people to die from lack of healthcare. The State of Maine, where I now live, with its Tea Party Governor, is one of those states. Our healthcare system until the Affordable Healthcare Act went into effect was one of the worst in the world (measured by such things as longevity, percentage of people covered, infant mortality, etc.), and without the conspicuous success of "Obamacare" would soon have cost 25% of our annual GDP. This would have happened especially rapidly if the Republicans had won the Presidency and both houses of Congress in the elections of 2012. Happily, Barack Obama was re-elected, and the Democrats kept a majority in the Senate for the next two years, though the Republicans continued to control the House of Representatives. The latter have voted over and over again to repeal the Affordable Health Care Act. Their announced goal still remains to repeal it, to reduce Social Security or phase it out, to cut back or eliminate Federal contributions to Medicaid, and to turn Medicare into a voucher system. These acts would increase by about sixty million persons the number of Americans without healthcare insurance, not to speak of raising catastrophically healthcare costs for individuals. Meanwhile, any fool can see that single payer universal health care, such as all other "first world" countries have, is the way to go both for reduced costs and for improved healthcare.

I have spoken of "lying propaganda." Perhaps it might be better to say that, as Guy Debord and Jean Baudrillard affirmed, in cyberspace the distinction between truth and lie has vanished in a triumph of "the society of spectacle" (Debord 1973) or of "simulacra" (Baudrillard 2013). I shall return in a later section of this essay to Debord and Baudrillard. One of

Mitt Romney's top advisers in the presidential campaign of the Fall, 2012 asserted that it did not matter whether or not what Romney proclaimed in his campaign speeches was true or not, so long as it was believed. Romney's defeat was to some degree a triumph for the survival of a belief that the distinction between truth and lie is still relevant.

4) We are still mired in all sorts of ways in the Middle East, in spite of the troop withdrawals that Obama promised and has more or less carried out, over the prostrate bodies of Republican hawks. If the Republicans win the Presidency and both houses of Congress in 2016 we are almost certain to be at war big time in the Middle East again in a matter of months. The wars in Iraq and Afghanistan cost trillions of dollars, hundreds of thousands of lives, mostly natives of the countries we have invaded, but thousands of Americans too. Hundreds of thousands of our returned soldiers suffer from Post Traumatic Stress Disorder.

These wars have contributed to a new American xenophobia and to the serious erosion of our constitutional liberties here in the United States: suspension of habeas corpus, illegal surveillance, for example of email and cell-phones, including following GPS locators in cell-phones that cannot be turned off just by turning off the phone, murder by drone aircraft that are fantastic surveillance machines as well as carriers of deadly missiles. Just yesterday (April 23, 2015) came the announcement an "unfortunate" drone strike in Pakistan in January that killed two hostages, an American and an Italian, held by Al Qaeda, along with several Al Qaeda members (Baker 2015).

Big Brother is watching us in America these days with a vengeance. This police-state surveillance in the United States has come about since 9/11 in the name of the so-called "War on Terror." The media barrage associated with that "War" is calculated to keep citizens in a constant state of terror. That foreign-looking man or woman with a package next to us in the supermarket or at the airport may well be a terrorist. We move away from her or him in case a hidden suicide bomb goes off.

5) New global telecommunication devices are putting people in touch with one another worldwide, thereby weakening local communities. Computer programs have played a big role in making possible the global

financial meltdown. Though the Internet makes a lot of useful truth available by a mouse-click, as I shall later specify, it is also the means of promulgating a lot of lies, for example in the deliberate use on a large scale of Facebook and Twitter to distribute false political propaganda.

Both drone warfare and cyberwarfare would be impossible without the digital revolution. The Pentagon has a Cyber Command to oversee retaliation and protection against attacks by cyberweapons on our government computer systems, a well as to carry out preemptive strikes such as the one we used some years ago to disable Iran's nuclear enrichment facilities (Sanger and Shanker, 2013). The United States uses cyberwarfare big time, for example to hack Chinese government computer servers or the private files of German Chancellor Merkel.

The new digital devices—computers, iPhones, iPads, Facebook, Twitter, video games, and the like—are rapidly diminishing the role literature plays in most people's lives. That is a big loss, but not the end of civilization, any more than was the shift from manuscript culture to print culture. There is no use, in any case, lamenting what is happening to the social role of literature. It will go on happening anyway. People these days play video games or watch films on Netflix or surf the net instead of reading Jane Austen, George Eliot, or Henry James in printed books or online, where huge amounts of print literature are now available. Our universities are, like glaciers worldwide, also in meltdown mode, especially the humanities.

Though the Internet has been praised for all sorts of reasons, it may well be causing with amazing rapidity a disquieting mutation in what it means to be human. To be human these days more and more means being plugged into some digital device or another, as I am at this moment plugged into my computer in order to write or revise this essay. Who can doubt that cell phones, computers, and other digital devices facilitated the Arab Spring and other such insurrections around the world as well as being essential to the rise to power of ISIS, for example by the recruiting for ISIS of American young people, both male and female?

The amazingly rapid development of MOOCs, "Massive Open Online Courses" is making "face to face" university and college teaching more and more a thing of the past. A 2013 article in the *New York Times* reported that the co-founders of Coursera, two computer science professors at Stanford, "watched with amazement as enrollment passed two million last month, with 70,000 new students a week signing up for over 200 courses, including Human-Computer Interaction, Songwriting and Gamification, taught by faculty members at the company's partners, 33 elite universities" (Lewin, 2013). Training in "Gamification" is just what we need.

In 2013, as I was revising this essay yet once more by way of the marvelous resources of Word for Mac, another essay in praise of MOOCs appeared in the *New York Times*. This was an enthusiastic Op-Ed piece by Thomas Friedman, "Revolution Hits the Universities" (Friedman, 2013). Many similar essays have appeared since then. The 263 blogs responding to Friedman's essay are interesting as indications of what people thought about MOOCs in 2013 and probably still do often think. Most, but by no means all, of the bloggers express more or less cogent opposition to MOOCs. The opposition is primarily by university and college professors and administrators. This is not surprising, since MOOCs not only appear possibly able to put colleges and universities out of business (including for profit ones like Phoenix University), but also appear possibly able to make professors and even adjuncts obsolete. Why pay $30,000 to $50,000 a year to attend the University of California at Irvine, or Oberlin, or Harvard, when I can take the same courses more or less for no cost on my computer, taught before a video camera by distinguished professors at elite universities?

According to a National Public Radio Interview by Ira Flatow in 2010 of an associate professor of neurology, physiology, and psychiatry at the University of California, San Francisco, Adam Gazzaley, and of a professor of communication at Stanford, Clifford Nass, research has shown that the brains of those who spend a lot of time using iPhones, iPads, checking and sending email, doing a lot of tweeting and texting, listening to MP3 music, all at the same time, are rewired for short attention spans and for multitasking, as opposed to the rather different wiring necessary for the long-term single-minded concentration required, say, to read *Middlemarch* (Flatow; Gazzaley; Nass, 2010) A new kind of digital dexterity and hand-eye coordination, moreover, is needed to use an iPhone, or to play a video game, or to manipulate a mouse. Those changes are what I mean by a mutation in what it means to be human. Our brains and bodies are becoming different as we are plugged in more and more of the time to the new digital "language systems."

The recent financial meltdown was facilitated by the computerizing of stock trading and other financial transactions. Digital machines, for example, automatically do the work of "high frequency trading," buying and then selling stocks and bonds in a few microseconds for each trade, far faster than any human decision could be made. Attempts to restrain this by charging a minuscule fee for each trade have of course been met with fierce opposition. Ways to rig the market have evolved much faster than any effective regulation of computerized financial trading. On May 6, 2010, a "Flash Crash" dropped the Dow Jones industrial average 1,000 points in a few hours. On August 1, 2012, a "rogue algorithm" in a stockbroker's program for automatically buying and selling stocks suddenly made millions of trades when the market opened, "spreading turmoil across Wall Street." A specialist on markets, Patrick Healy, said: "the machines have taken over, right?" (Popper, 2012). Of course computer nerds wrote those programs and algorithms, including the ones with inadvertent glitches. Nevertheless, once such programs are embedded in a powerful server they just go on working on their own, like Kafka's robotic Odradek. Such programs still sometimes have a propensity to self-destruct hidden as "bugs" within them. These may freeze the system or cause other trouble when someone tries to use the program in a way not foreseen by the programmer.

Just as I am making this revision, that is, five years after the "Flash Crash" event, a little-known British futures trader, Navinder Singh Sarao, whose "spoofing" (that is putting in orders to raise the price of a financial asset and then cancelling the order) authorities hold at least partly responsible for the Flash Crash, has been arrested (Popper and Anderson, 2015). James Weatherall, in an essay in *Nautilus* for April 23, 2015, "Why the Flash Crash Really Matters," argues, however, that "The Flash Crash reveals that we need a fundamentally different understanding of how modern financial markets work. We believe [he means "I believe"] that it shows us [Who is included in this "us"?] that markets are governed by the same principle as earthquakes and avalanches: self-organized criticality" (Weatherall, 2015). "Criticality" means, as the essay shows, that the vast sign system making up financial markets is an example of what de Man calls "the linguistics of literarity." Another way to put this is to say that chaos theory or catastrophe theory or "criticality" (the butterfly wing flap in Guatamala brings about tornadoes in Kansas) operates in the huge digital sign system of the financial markets too.

My goal in this first section has been not only to identify features of these "bad times," but also to show that all these features, including climate change, have an inextricable involvement in language, most often with the power of ideological lies. I now turn to Paul de Man's "The Resistance to Theory" to try to identify exactly what de Man says about ideology in that essay.

2. If You Want to Lie, Digitize

The frightening realities I have sketched out are well known to most Americans. How can we justify spending time, in such a dire situation, with something seemingly so marginal or even so trivial as literature or literary theory? My colleagues, Claire Colebrook and Tom Cohen, have, in their essays in the companion volume to this one (Cohen, Colebrook, Miller, 2012), brilliantly confronted this question. They did this with special attention to how useful or even indispensable it is to read Paul de Man carefully today, no easy task. Their essays for this volume continue that work. Colebrook has suggested that de Man may be *the* theorist for the twenty-first century. I agree. Just how might that be the case?

You will note that the five catastrophes I have listed are all brought about by people's propensity to believe lies. One scandal to cognition is the difficulty in understanding why this propensity is so deeply rooted. How in the world can so many people be brought, for example, to vote against their own self-interest, to shoot themselves in the foot, so to speak? That seems an unfathomable mystery. We have done nothing to mitigate climate change because so many people have, against all the evidence, believed lies that claim it is a hoax. "How could 'Mother Nature' or a Good God allow such a thing to happen?" Governments in Europe and America have believed the lie that austerity measures and lowering taxes on the rich (the "trickle-down theory) will reduce the deficit and create many good jobs, whereas Economics 101 and recent Eurozone history show that this is false. Bank deregulation depended on accepting the lie that the financial system could be trusted to regulate itself, whereas the banks and other financial institutions just used deregulation to commit the widespread fraud that led to the mortgage crisis and global recession. Resistance to adopting the obvious solution to the healthcare cost crisis, namely universal single payer healthcare, such as all other "first world" countries have in some form, has been fueled by endlessly reiterated lies in the media about how such healthcare would "pull the plug on grandma," and so on. The catastrophic wars in Iraq and Afghanistan were based on the lies that Saddam Hussein had weapons of mass destruction, that both countries were centers of Al Qaeda, and that both countries would welcome us with open arms as liberators bringing them the Western-style democracy they supposedly covet. The latter was especially Dick Cheney's iterated lie. Twitter, Facebook, and other new "social media" harnessed to the Internet by iPhones and other such gadgets have given enormous new scope to the ability to tell lies and get them widely believed. These devices have even eroded for many the distinction between truth and lie.

3. What is Ideology for de Man?

De Man's name for lies that are believed is "ideology." This returns me to my epigraph. Just what is de Man really saying about ideology? I want to identify that as exactly as possible both for my citation and for "The Resistance to Theory" as a whole—if it is a consistent whole, that is, which by no means goes without saying. Just before the passage cited as my epigraph, de Man has been talking about the aboriginal human error of taking language as referentially valid, as having "authority as a model for natural or phenomenal cognition" (de Man, 1986, 11). To put this another way, ideology means believing that lies are true, for example that just because some politician or talk show celebrity says Barack Obama was not born in the United States and is secretly a Moslem socialist, those false fictions (one might even call them "literature") are true. "Literature," says de Man, "is fiction not because it somehow refuses to acknowledge 'reality,' but because it is not *a priori* certain that literature is a reliable source of information about anything but its own language" (de Man, 1986, 11).

De Man goes on in the next paragraph to say that "it would be unfortunate, for example, to confuse the materiality of the signifier with the materiality of what it signifies" (de Man, 1986, 11). Hardly anyone will be likely to do that with words about material phenomena, but with fictional narratives about history, politics, or life stories we are much more likely to be taken in: "No one in his [sic!] right mind will try to grow grapes by the luminosity of the word 'day,' but it is very difficult not to conceive the pattern of one's past and future existence as in accordance with temporal and spatial schemes that belong to fictional narratives and not to the world" (de Man, 1986, 11). We in imagination make our lives confirm to fictional conventions, particularly those of "romance": "poor little boy with no opportunities makes good through hard work and luck"; "pretty but naïve teenage girl meets Prince Charming in disguise, ultimately marries him and lives happily ever after."

"This does not mean," de Man goes on to say, "that fictional narratives are not part of the world and of reality; their impact upon the world may well be all too strong for comfort" (de Man, 1986, 11). Fictional narratives make things happen if we believe them and act on that belief. They have become felicitous speech acts, a way of doing things with words. A lot of Americans, happily not quite enough, voted for Mitt Romney in the 2012 US presidential election. Obama's plurality was not all that great. Romney might well have been elected. A Republican may well be elected President in 2016, a terrifying thought, since all the likely candidates are in one way or another Tea Party zombies who have already been bought by corporations and the Koch brothers or Adelson. Further examples from our recent history would include the invasion of Iraq on the basis of that lie about Saddam Hussein having weapons of mass destruction, or the delay until it is too late to do anything about CO2 emissions because so many people have been persuaded to believe climate change predictions are a conspiracy perpetrated by power-mad conspiratorial scientists, or the belief that the Affordable Healthcare Act is a "government take-over" that will make our healthcare worse and more costly, whereas just the reverse has happily turned out to be the case. That does not keep the Republicans from going on doing their best to repeal Obamacare. HMO's and pharmaceutical companies have bought them through big covert contributions.

The passage from "The Resistance to Theory" I cited as an epigraph follows just after the passages I have quoted and discussed above. Let me look at my epigraph a little more closely. What does it really say? It is the most eloquent assertion by de Man of the specifically political and social value of literary theory. Literary theory is study of "the linguistics of literariness," by which de Man means, as the rest of his essay and his work generally confirm, study of the role of rhetoric (meaning tropes) as disruptive of grammatical and logical meaning in texts generally. Grammar, logic, and rhetoric are the elements of the medieval *trivium*, the basic divisions of language study. De Man consistently discriminates among these three elements of language. That de Man means by "the linguistics of literariness" something that goes beyond printed literature in the sense of poems, plays, and novels to include features of texts in general is confirmed by a passage later in the essay: "The resistance to theory is a resistance to the rhetorical or tropological dimension of language, a dimension which is perhaps more explicitly in the foreground in literature (broadly conceived) than in other verbal manifestations or—to be somewhat less vague—which can be revealed in any verbal event when it is read textually" (de Man 1986, 17). Elsewhere in "The Resistance to Theory" de Man includes graphic images along with strictly verbal events (words on the page) as requiring "textual" reading, as I shall show later on in this essay.

Tropological displacements are, de Man makes clear, the basic instruments of ideological fictions or, to put it more bluntly, the means of telling lies that people are led to believe are true and on the basis of which they then act disastrously. "What we call ideology is precisely the confusion of linguistic with natural reality, of reference with phenomenalism." Note that de Man says it is a matter of "calling" it ideology, of arbitrarily assigning it a name, not a matter of using the proper name of some ideal essence. Calling it ideology is a speech act, not a constative truth. This is partly because calling it ideology brings in the history of a loaded word, from the late eighteenth-century French "ideologues" on to Marx and Nietzsche and then to present-day ideology-critique in Lyotard, Deleuze and Guattari, or Derrida.

"Phenomenalism" is a word de Man chooses carefully to indicate that he is not talking about "reality" as it is in itself, but as it appears to our limited senses. Ideology is a confusion of linguistic with natural reality as the latter appears to our eyes, ears, taste, and touch. In my terms here, ideology means taking a lie, a fictional construction, as true in reality so-defined. Just because a plausible story, complete with aerial photographs and maps, is told about Saddam Hussein having weapons of mass destruction, we believe that he really does have them. We then invade Iraq on the basis of that false assertion. We vote to elect a congressman or congresswoman who will do everything possible to repeal the Affordable Healthcare Act because we believe the lying propaganda against that Act. We support deregulation of banks and the financial system because we have been persuaded to believe the lie that the financial system can be safely trusted to regulate itself honestly. A lie, as I have said, becomes a felicitous speech act if it is believed and acted upon. We vote for Mitt Romney because we have been persuaded to believe the lie that lowering taxes on the rich and on big corporations, while firing teachers, halting infrastructure projects, gutting Medicare, Medicaid, and Social Security, will magically produce millions of new jobs. All these cases are examples of trying to grow grapes by the luminosity of the word "day."

4. Why Study Literary Theory?

Well, how does studying literary theory or carefully reading Paul de Man help us in this self-destructive situation? Derrida compared our situation, notoriously, to having an autoimmune disease that turns the body's defensive antibodies against its own tissues and organs. De Man's next sentence in my epigraph (I cite it again) gives a bold, defiant, cheeky, and even counter-intuitive answer to the question of the utility of literary study: "It follows [from de Man's definition of ideology] that, more than any other mode of inquiry, including economics, the linguistics of literariness is a powerful and indispensable tool in the unmasking of ideological aberrations, as well as a determining factor in accounting for their occurrence. Those who reproach literary theory for being oblivious to social and historical (that is to say ideological) reality are merely stating their fear of having their own ideological mystifications exposed by the tool they are trying to discredit." Please note that de Man here rather casually makes the truly shocking assertion that "social and historical reality" are ideological through and through, that is, a matter of false language believed in as true.

The reference to Marx follows immediately. Just what does de Man mean by being good readers, as opposed to very poor readers, of Marx's *German Ideology*? What justifies bringing in Marx at this point? Answering those questions at all adequately would take another long essay or perhaps even a book. I have made a beginning with this in "The Working of Material Spirit" (Miller, 2013). Guessing what the reference to Marx means would involve, for example, deciding whether Derrida, in *Specters of Marx,* would have seemed to de Man a good reader of *The German Ideology.* Would de Man and Derrida have read what Marx says about the essence of ideology in the same way? In *Specters of Marx* Derrida, following Marx, traces all ideological aberrations back to the notion of the Man-God, the material Incarnation of spirit: "Is not the Christic moment, and within it the Eucharistic instant, the hyperbole of *acharnement* ["opinionated ardor," from *charnière,* hinge, point of junction] itself? If every specter, as we have amply seen, is distinguished from spirit by an incorporation, by the phenomenal form of a quasi-incarnation, then Christ is the most spectral of specters" (Derrida 1994, 144; Derrida 1993, 229). I suspect de Man and Derrida would read Marx in *almost* the same way. I say "almost" because Derrida was tempted much more than was de Man by the lure of some "wholly other" and by what he called "a messianic without messianism" (Derrida 1994, 59; Derrida 1993, 102). This formulation attempts to avoid any belief in an actual Messiah to come, any belief in the Incarnation. It is still, nevertheless, tempted by the messianic. De Man would not, I imagine, have altogether agreed with Derrida's formulations. Nevertheless, a reading of Marx by de Man, I suspect, if he had done one, would have stressed, as Derrida does explicitly in *Specters of Marx,* that for Marx the source of all ideological delusions is the spectral and baseless mystification of religious belief. More specifically, all ideological aberrations are modeled on belief in the Incarnation, that is, belief that in Christ the Word spirit and matter were conjoined. Jesus's body incarnated God. Commodity festishism, for example, as analyzed by Marx in *Capital,* falsely believes, in way analogous to belief in the Incarnation, that a given commodity embodies value. This is like trying to grow grapes by the luminosity of the word "day." It is, that is, belief that the word "day" incarnates a performative power over nature. A later section of this essay will analyze one example of commodity fetishism in a recent ad from *Wired Magazine.*

De Man is advocating using the linguistics of literariness, or what he also calls "rhetorical reading" (the sort of readings de Man made of Nietzsche, Rousseau, Benjamin, Baudelaire, and many others), to read the ideological lies that I began by identifying and that have got us by way of sleepwalks into our present catastrophic situation. Rhetorical readings of talk shows and news as promulgated by various media, of television ads, of films, of politicians' statements, of political advertising, or of Supreme Court decisions will work better than any other form of analysis, de Man in effect promises, to "unmask ideological aberrations," to tear off their masks and show the sinister face or impersonal lack of face that lies beneath. The linguistics of literariness, moreover, promises to "account for their occurrence," that is, to explain how these disastrous self-destructive aberrations come about in the first place. That explanation would go by way of readings of the role of tropological displacements in making falsehoods seem plausible.

It would seem to follow from this redoubtable power of "unmasking" linguistic aberrations that once we understand them we will be cured, enlightened. We will also act henceforth on that enlightenment. In a seemingly analogous way, Freud promulgated the "talking cure" that would through language about language and other signs, such as dream images, cure psychosomatic maladies. These arise, Freud believed, if we are male, from unconsciously wanting to have sex with our mothers and therefore wanting to kill our fathers, or, if we are female, from hysteria that arises from a tropological similarity between a childhood sexual abuse by the father not understood as such at the time and some much later similar event that may not in itself be traumatic.

The reader will no doubt have already observed that de Man's key word in his essay, "resistance," is borrowed, surprisingly, from psychoanalysis. Freud used the term "resistance" (to oversimplify the issue a bit) to refer to patients' blocking memories from conscious memory, just as those under the spell of an ideological aberration resist being cured of their mistake. I say "surprisingly" because de Man was generally hostile to psychoanalysis. "Resistance," for de Man in "The Resistance to Theory," however, is a linguistic phenomenon rather than a feature of consciousness in its relation to memories suppressed in the unconscious. De Man has tropologically appropriated the term for his own uses.

Powerful resonances with the psychoanalytical usage are nevertheless still present. If we practice de Man's talking cure, de Man seemingly implies, we will no longer be ideologically mystified. We will then vote correctly, support universal single-payer healthcare, endorse strong regulation of banks, abandon our imperialistic wars, mitigate climate change, deal with the deficit by taxing the rich, and so on. We will, in short, support all those changes Barack Obama endorsed in his second inaugural address. De Man's essays can, it would appear, be seen as exemplifications of the way rhetorical readings work as salutary enlightenment. After such readings we know "what Benjamin really says," or what Kant, Hegel, Rousseau, or Nietzsche really say. That turns out usually to be as different from the traditional interpretations of those authors as believing that Hussein had weapons of mass destruction is from the fact that he did not.

5. What Does "The Resistance to Theory" Really Say?

Whether this would really happen and whether de Man is really promising that rhetorical readings or wisdom about the linguistics of literariness will make us all vote right is a big question. It does not go without saying that de Man is making that utopian promise. I shall eventually return to this issue, but need to work carefully toward it. Just after the passage about the superiority of the linguistics of literariness over every other mode of ideology critique, de Man explains "what it is about literary theory that is so threatening that it provokes such strong resistances and attacks": "It upsets rooted ideologies by revealing the mechanics of their workings; it goes against a powerful philosophical tradition of which aesthetics is a prominent part; it upsets the established canon of literary works and blurs the borderlines between literary and non-literary discourse. By implication, it may also reveal the links between ideologies and philosophy" (de Man 1986, 11). De Man's prime target in his late work, you will remember, was what he called "Aesthetic Ideology." That helps account for the word "aesthetics" in this passage.

The statement just cited certainly goes far to explain the resistance to theory. That resistance is analogous to our resistance to what scientists say about so-called "anthropogenic" climate change. (See Cohen's essay in this book for an identification of the big problems with the word "anthropogenic," as with "anthropomorphic," and "Anthropocene." A big part of the problem is the prefix "anthropo-" in all three words. "Anthropo-" begs important questions by buying into the Greek definition of "man" as male, reasonable, and Greek, that is, the prefix is intrinsically exclusionary, as the Greeks excluded women and all non-Greek others, the "barbarians," from the category of "man.") We want to remain safe within the warm cocoon of our ideological aberrations, whereas literary theory shows they are mistakes. It unmasks them. The reception of de Man's own work is a spectacular example of this resistance to being woken up from our illusions. His work always had an amazing ability to send readers or listeners up the wall, as for example those who heard de Man at Yale and Cornell present his lecture on Walter Benjamin's "The Task of the Translator."

"The Resistance to Theory" itself is another notable example. The essay was commissioned by the Modern Language Association of America as part of a book entitled *Introduction to Scholarship in Modern Languages and Literatures.* De Man was asked to write the essay on literary theory. The committee in charge of this book had an opportunity to publish one of the most brilliant and challenging essays in twentieth-century literary theory. They blew it big time by rejecting the essay. That was surely a mistake. It was a revealing error, however, since it was perhaps based, whether the committee members consciously knew it or not, on a feeling that their ideological convictions were deeply threatened by the essay. De Man's account, near the beginning of the essay as it was ultimately published in *Yale French Studies* in 1982, is devastating in its gentle irony: "I found it difficult to live up, in minimal good faith, to the requirements of this program [as indicated in the title of the MLA book cited above] and could only try to explain, as concisely as possible, why the main theoretical interest of literary theory consists in the impossibility of its definition. The Committee rightly judged that this was an inauspicious way to achieve the pedagogical objectives of the volume and commissioned another article. I thought their decision altogether justified, as well as interesting in its implications for the teaching of literature" (de Man 1986, 3). The *Yale French Studies* issue that finally published the essay was about teaching literature. De Man's "interesting in its implications for the teaching of literature" suggests that if literary theory cannot be defined that may mean that it is also impossible to "teach literature" without remaining imperturbably within those aberrational "rooted ideologies" of which literary theory reveals the hidden mechanics. I shall return later in this essay to what is implied by asserting that ideologies work "mechanically."

So far so good. It would appear that literary theory is good because it demystifies dangerous ideological mistakes. Who could be against that, once you come to understand what is at stake? Matters are not so simple for de Man, however, as the rest of his essay shows, as does that ominous formulation just cited about the impossibility of defining literary theory. Surely we ought to be able to figure out what theory is and define it! The rest of the essay explains why that is not the case.

De Man begins anew, after the formulations near the beginning I have been analyzing, by saying, quite surprisingly, that the resistance to theory when seen as an attempt to protect rooted ideologies from exposure is relatively uninteresting. It is of merely local and historical importance. It is something belonging to our own parochial time and place, such as the room at the MLA headquarters where that committee met to reject de Man's essay.

What interests de Man more, he says, is something universal, something that is true of the resistance to theory at any time or place. The reader of de Man's essay must follow its intricacies for herself or himself. I shall continue now, however, by investigating the implications of a series of sentences that take the form "A is B." These copulative sentences form the backbone of the argumentation developed in "The Resistance to Theory." Such copulas punctuate the essay at intervals like a recurrent modulated motif. They could be run together in a string that would say "A is B is C is D," and so on. Are these sentences logical, definitional, positional, or axiomatic? Are they examples of that seventeenth-century *more geometrico* de Man discusses in this essay, as in the definition of a triangle: "A triangle is ..." Or could the sentences be (false) tropological equivalences, since, like a metaphor, they assert the identity of two entities that must be different, one would think, since they have different names. Aristotle's maritime example in the *Poetics* of a metaphor, "The ship ploughs the waves," can be re-phrased as "A ship is a plough." Anybody can see that a ship is not a plough. Interesting that so often examples of metaphor given by theorists involve ships, that is, a means of transport. That is after all what "metaphor" etymologically means, "carry over." A metaphor gets you from here to there.

It makes a lot of difference which way de Man's "is" sentences are taken. I do not think this can be decided. This uncertainty indicates that even a thoroughly enlightened theorist of tropes like de Man cannot avoid repeating a version of the deplorable uncertainty he exposes in others. An example is his analysis in this essay (discussed below) of the undecidability of the words "of" and "fall" in Keats's title, *The Fall of Hyperion.* It seems as if just knowing does not prevent you, by a fatality inherent in language, from erring, from repeating the undecidability you unmask. Knowing, pace Freud, does not cure resistance to theory because theory is itself the resistance to theory. I shall return to this disquieting formulation later on.

One thing is clear, nevertheless. De Man's "is" sentences, whether we decide they are tropes or propositions, are conspicuously scandalous and counterintuitive. I believe de Man deliberately makes them that way, perhaps because such bald statements are more likely to penetrate our ideological prejudices or at least arouse our conscious resistance. De Man's "is" sentences are increasingly illogical, as the sequence proceeds. It appears that what de Man wants to say can only be said illogically, by way of paradoxical equivalences asserted between entities that are manifestly different. Here are the sentences in question, in their sequence in the essay, with "is" italicized in each case:

> What we call ideology *is* precisely the confusion of linguistic with natural reality, of reference with phenomenalism (de Man 1986, 11).
>
> The resistance to theory *is* a resistance to the use of language about language. It *is* therefore a resistance to language itself or to the possibility that language contains factors or functions that cannot be reduced to intuition (de Man 1986, 12-13).

> It turns out that the resistance to theory *is* in fact a resistance to reading .... (de Man 1986, 15).
>
> The resistance to theory *is* a resistance to the rhetorical or tropological dimension of language .... (de Man 1986, 17).
>
> The resistance to theory which, as we saw, *is* a resistance to reading .... (de Man 1986, 17-18).
>
> Nothing can overcome the resistance to theory since theory *is* [in this case de Man himself italicizes the "*is*"] itself this resistance (de Man, 1986, 19).

What does this sequence mean? What is "reading" doing in the sequence? Just what *is* reading for de Man? What in the world can de Man mean by saying that the resistance to theory does not come from outside, from those whose rooted ideologies are threatened by theory, but is intrinsic to theory itself? Theory *is* the resistance to theory! That is a scandalous "is" sentence if there ever were one. What implications does that have for the apparent claim near the beginning of the essay that theory would liberate us from ideological mistakes by showing us how they work? How could theory do that if we cannot even define it and if the resistance to theory is intrinsic to theory itself? That would seem to make theory pretty feeble and useless, a broken reed, as opposed to being, as de Man claims, "a powerful and indispensable tool in the unmasking of ideological aberrations."

I cannot pretend, much as I would like to do so, in a few sentences to give clear literal answers, *more geometrico*, to these questions, much less to follow in detail the intricate argumentation of de Man's essay, along with the "logic" of his casual putdowns along the way of all the then-influential theoretical schools: new criticism, structuralism, semiotics, phenomenological criticism, criticism of consciousness, reader response criticism, and criticism using speech act theory. All are mystified, de Man argues, because they resist rhetorical reading and resist study of the linguistics of literariness. A beginning of a reading of "The Resistance to Theory" can be made, however, in relation to my fundamental question: Why do so many people believe lies these days and remain mystified by aberrational ideologies? My sketch of a reading will lead to somewhat dismaying tentative results.

6. Two-Handedness as Sleight of Hand

It looks to me as if de Man takes away with one hand what he gives with the other. On the one hand, he unreservedly praises the theory of rhetorical reading, that is, the linguistics of literariness, for being a powerful tool, more powerful even than economics, for "unmasking" rooted ideologies. On the other hand, when he turns to the intrinsic study of theory, he finds that the most important resistance to it does not come from outside theory, that is, as embodied in some people's "fear at having their own ideological mystifications exposed by the tool they're trying to discredit" (de Man 1986, 11). The true resistance to theory comes from inside theory itself. The assertion that the resistance to theory is intrinsic to theory is the amazing climax of the "'is' sentences" I have listed: "Nothing can overcome the resistance to theory since theory *is* itself this resistance."

How in the world can that be? Theory would appear to give a clear model for salutary rhetorical reading of any text whatsoever, including political speech and television or magazine advertising, etc. The answer lies in what happens when the program of rhetorical reading is carried out in a given case. What rhetorical reading is we know or think we know. It is attention to the way the tropological dimension of any discourse interferes with its statement of a clear logical meaning. "Poetics" (*die Art des Meinens*, the way meanings are expressed) interferes with "hermeneutics" (*das Gemeinte*, what is meant). Here is what de Man says about this interference in "Conclusions: Walter Benjamin's 'The Task of the Translator'":

> When you do hermeneutics, you are concerned with the meaning of the work; when you do poetics, you are concerned with the stylistics or with the description of the way in which a work means. The question is whether these two are complementary, whether you can cover the full work by doing hermeneutics and poetics at the same time. The experience of trying to do this shows that it is not the case. When one tries to achieve this complementarity, the poetics always drops out, and what one always does is hermeneutics. One is so attracted by problems of meaning that it is impossible to do hermeneutics and poetics at the same time. From the moment you start

> to get involved with problems of meaning, as I unfortunately tend to do, forget about the poetics. The two are not complementary, the two may be mutually exclusive in a certain way, and that is part of the problem which Benjamin states, a purely linguistic problem (de Man 1986, 87).

The play of pronouns here ("one," "you," "I") implies that I cannot avoid repeating, for example in this essay, the betrayal de Man names and with rueful irony confesses to performing. I do hermeneutics at the expense of poetics, as when I ask, "What does de Man really say in this essay?" That question implies that what he says can be clearly identified and paraphrased, that it is not interfered with by the way de Man says it.

Theory is resistance to reading, as de Man implies, though he does not say so in so many words. This formulation would add one more "is" sentence to those de Man enunciates. Theory is resistance to reading apparently because theory pretends to foresee clearly the results of reading (demystification of aberrational acts of taking metaphors literally), whereas reading itself, rhetorical reading, is unpredictable. You never know beforehand just what you are going to find in a given text. Each genuine reading is, consequently, sui generis. It is not reducible to the application of a formula that knows what it is going to find. De Man's teaching was a conspicuous example of this. Even when I thought I knew intimately already the text he was going to read in a given seminar and even when I thought I could guess what he was going to say on the basis of what I already knew of his work and of the work he was going to discuss, I was always surprised. He always presented a new way of reading that particular text, a way I had not foreseen. One result of theory's resistance to reading is that clear theory (after all "theory" etymologically means "clear seeing," as in "theatrical"), when it becomes a model or tool for reading, leads to readings that become more and more opaque. Ultimately it culminates in (I am sorry, but I must use this word) "undecidability." Reading and theory are no more compatible than are hermeneutics and poetics. Reading leads to the disqualification or severe modification of theory. De Man elsewhere calls this event an encounter with "the impossibility of reading," which "should not be taken too lightly" (de Man 1979, 245). An example is the impossibility of deciding whether de Man's uses of "is" are logical or tropological. We urgently need to know, but we cannot know for sure. My own serene assumption at the beginning that I can know a lie when I see one is beginning, by the way, to look increasingly problematic.

The big (and only detailed) example in "The Resistance to Theory" of this dismaying outcome of following out the injunction to use rhetorical theory as a tool for unmasking ideologies is the paragraph about the meaning of Keats's titles for his two unfinished poems or versions of the same poem, *Hyperion* and *The Fall of Hyperion*. Only de Man could have written this paragraph. It is vintage de Man. He is inimitable. Don't try this at home. You must do rhetorical reading, if you want to do it, in your own unique way.

Beginning with the impossibility of deciding whether the "of" in *The Fall of Hyperion* is a subjective or objective genitive ("Hyperion's Fall" or "Hyperion Falling"), de Man makes a rapid series of moves. "Both readings," he says, "are grammatically correct, but it is impossible to decide from the context (the ensuing narrative) which version is the right one. The narrative context suits neither and both at the same time, and one is tempted to suggest that that the fact that Keats was unable to complete either version manifests the impossibility, for him as for us, of reading his own title" (de Man 1986, 16). This leads, by way of an expansion of the proper name "Hyperion" to potentially name Apollo, Keats, and, ultimately, the reader, to a clear statement of the alternatives between which we cannot decide. This happens, in part, through two uses of a characteristic de Manian locution for saying yes and no simultaneously: "manifestly, yes, but not quite." This is a deliberate breaking of the logical law of the excluded middle. A variant, "This is indeed the case, to some extent; but not quite," appears on the next page (de Man 1986, 17). "But," asks de Man, "are we then telling the story of the failure of the first text as the success of the second, the Fall of *Hyperion* as the Triumph of *The Fall of Hyperion*? Manifestly, yes, but not quite, since the second text also fails to be concluded. Or are we telling the story of why all texts, as texts, can always be said to be falling? Manifestly yes, but not quite either, since the story of the fall of the first version, as told in the second, applies to the first version only and could not legitimately be read as meaning also the fall of *The Fall of Hyperion*" (de Man 1986, 16).

The paragraph concludes by drawing some dismaying conclusions about this miniature exercise in rhetorical reading. Crucial to what de Man says in these sentences is the assumption that reading literature is not innocent. It has consequences. Reading does something to us. "The difference between the two readings," says de Man, "is itself structured like a trope. And it matters a great deal how we read the title, as an exercise not only in semantics, but in what the text actually does to us. Faced with the ineluctable necessity to come to a decision, no grammatical or logical analysis can help us out. Just as Keats had to break off his narrative, the reader has to break off his understanding at the very moment when he is most directly engaged and summoned by the text" (de Man 1986, 16-17).

On the one hand, you must obey the injunction to read rhetorically, that is, to follow out the interference of the rhetoric of tropes in the working of grammar and logic, of poetics as interfering with hermeneutics. It is an ineluctable summons. Your life, the future of your country, of the planet, and of the human race depend on doing such readings. On the other hand, it turns out that when you pursue this program in all good faith you find you cannot read in the specific sense of achieving clear and unequivocal knowledge of what a given text means. Rhetorical reading always leads you step by step to some undecidability, to some non-understanding. That is what de Man means by saying reading is impossible and why he says, at the conclusion of "Allegory of Reading (*Profession de foi*)," that "the impossibility of reading should not be taken too lightly" (de Man 1979, 245).

"The Resistance to Theory" ends with a dense paragraph that draws conclusions and generalizations from what de Man has said so far. He begins the paragraph by saying that we ought to do rhetorical readings because each is a salutary "methodological undoing of the grammatical construct and, in its systematic disarticulation of the *trivium*, will be theoretically sound as well as effective [effective presumably as the unmasking of ideological mystifications]. Technically correct rhetorical readings may be boring, monotonous, predictable and unpleasant, but they are irrefutable" (de Man 1986, 19). Charges of being boring, etc. were often made against such readings in those far-off days when they were still being done. De Man's phrasing is characteristically ironic. His readings of "fall" and "of" in Keats are anything but boring and predictable, though they may be exceedingly annoying and threatening to what we thought we could take for granted as achieved knowledge about Keats's poems. If the MLA refused to publish "The Resistance to Theory," Oxford University Press rejected *Allegories of Reading,* arguably de Man's most important work, though Oxford had published *Blindness and Insight* (1971). British Rousseau scholars that Oxford felt obliged to consult turned thumbs down on *Allegories of Reading.*

Deeply disquieting formulations such as commonly end de Man's essays begin to appear in "The Resistance to Theory" just after the ironic sentences about how technically correct rhetorical readings are boring and predictable but irrefutable. Such irrefutable readings, de Man says, are "totalitarian" (an ominous term), a form of non-knowledge, "unreliable, and "defective models of language's impossibility to be a model language." Here is the quite difficult sentence containing these elements: "They [irrefutable rhetorical readings] are also totalizing (and potentially totalitarian [meaning, I suppose, claiming complete sovereignty without any basis in some legitimate authority]) for since the structures and functions they expose do not lead to knowledge of an entity (such as language) but are an unreliable process of knowledge production that prevents all entities, including linguistic entities, from coming into discourse as such, they are indeed universals, consistently defective models of language's impossibility to be a model language" (de Man 1986, 19). I risk saying that de Man means here that those rhetorical readings, motivated by an impeccable theory, always led to some blank non-knowledge of just what we most need to know, an undecidability such as is manifested in de Man's reading of Keats's "fall" and "of." This non-knowledge means that theory always leads to some form of the unspeakable, to an experience of the impossibility of bringing some entity or other, including even language, "into discourse as such." They demonstrate that no such thing as a model theoretical language exists. Such defective models for reading are, nevertheless, though "always in theory," not in practice, "the most elastic theoretical model to end all models." They are this by containing "within their own defective selves all the other defective models of reading avoidance, referential, semiological, grammatical, performative, logical, or whatever. They are theory and not theory at the same time, the universal theory of the impossibility of theory" (de Man 1986, 19). Whew! One can see why that MLA committee was dismayed. The clear-seeing demanded by theory is impossible because the theory that contains all theories leads to non-knowledge whereas theory is supposed to be the opposite of non-knowledge, that is, a successful overcoming of ignorance. What would be needed is a clear theory of non-knowledge, and that is impossible.

The result is that though doing those boring but irrefutable rhetorical readings is eminently teachable, what is being taught is the resistance to reading and the resistance to theory that is built into theory itself, not something that comes to it from those idiots outside theory who instinctively hate theory as threatening to their ideological aberrations: "To the extent that they are theory, that is to say, teachable, generalizable and highly responsive to systematization, rhetorical readings, like the other kinds, still avoid and resist the reading they advocate. Nothing can overcome the resistance to theory since theory *is* itself this resistance" (de Man 1986, 19). Just because correct rhetorical readings are teachable does not necessarily mean it is a good thing to teach them. The reader may at this point remember a passage near the beginning of "The Resistance to Theory" where de Man, commenting on those who "denounce theory as an obstacle to scholarship and, consequently, to teaching," asserts that "it is better to fail in teaching what [according to such theory-haters] should not be taught than to succeed in teaching what is not true" (de Man 1986, 4). You will always fail in teaching the theory of the impossibility of theory, but what you can successfully teach should not be taught because it is a lie.

Oh dear, say I to myself at this moment in my investigation, there goes my hope to use, with de Man's approval, rhetorical readings as a tool in the fight against those five categories of lies I began by identifying as characterizing the twilight of the Anthropocene idols. Undecidability just does not seem a good tool for such demystification. De Man drives a final ironic nail into the coffin of my political hopes in the last three sentences of his essay. They end with a final example of his characteristically obscure and paradoxical formulations. The essay culminates in an explicit undecidability. Literary theory will go on flourishing because it thrives on its impossibility, its resistance to itself. Whether that is theory's triumph or its fall cannot be decided: "The loftier the aims and the better the methods of literary theory, the less possible it becomes. Yet literary theory is not in any danger of going under; it cannot help but flourish, and the more it is resisted, the more it flourishes, since the language it speaks is the language of self-resistance. What remains impossible to decide is whether this flourishing is a triumph or a fall" (de Man 1986, 20).

One can see why that MLA committee rejected de Man's essay. They were apparently smart enough to see that it threatens the entire enterprise of their book. That book was intended to help pedagogy by making available the most up-to-date information about the current state of scholarship in language and literature, including literary theory. De Man's essay does no such thing. It also puts in question the MLA's work generally of supporting scholarship and teaching in modern languages and literatures. Why should we go on trying to do something that is either "impossible" or wrong to do even when it is possible? As usual, de Man is unwilling to let his readers off the hook with some unequivocally positive or even just barely decidable conclusion.

In my naiveté, however, I still do not quite see how rhetorical readings of those five categories of lies I began by identifying might not make things better by unmasking them as the lies they are, even if those readings lead to aporias such as de Man identifies in Keats. The question is why politicians, media pundits, and voters persist in their ideological aberrations even when those delusions have been "unmasked" and clearly shown to be deeply mistaken and self-destructive.

De Man's essay gives the answer. The passage I began this essay by citing by no means says that doing readings in the light of the linguistics of literariness will cure you of making ideological mistakes. It just says that doing such readings will unmask such aberrations and account for their occurrence. Unmasking them and accounting for why they occur is not at all the same thing as curing people of their ideological aberrations. De Man's position is profoundly anti-Freudian. No "talking cure" for him. The argumentation of "The Resistance to Theory" as a whole, as I have tried to show, not to speak of de Man's work generally, claims, rather, just the opposite. It asserts not only that correct deconstructive readings do not lead to clear seeing but rather to undecidable aporias, for example the problems with "fall" in Keats. It also claims that such readings do not prevent the repetition of the mistakes they unmask and account for. Language itself programs the repetition of the aberration. You ought to do poetics and may even be strongly committed to doing so, but you always end up doing hermeneutics.

What is most distressing about de Man's analyses is that he shows that the effort of unmasking ideologies and of accounting for their occurrence in itself leads to a repetition of the linguistic aberrations that brought about the ideological mystifications in the first place. This is perhaps said most clearly near the end of "Allegory of Reading," de Man's essay in *Allegories of Reading* about Rousseau's *Profession de foi:* "Deconstructive readings can point out the unwarranted identifications achieved by substitution, but they are powerless to prevent their recurrence even in their own discourse, and to uncross, so to speak, the aberrant exchanges that have taken place. Their gesture merely reiterates the rhetorical defiguration that caused the error in the first place" (de Man 1979, 242). This citation comes three pages from the end of the essay on *The Profession de foi*. Its last sentence has already been quoted, but I come back to it now from another angle and in a different context: "One sees from this that the impossibility of reading should not be taken too lightly" (de Man 1979, 245). I take these two passages as confirmation that for de Man unmasking of ideological aberrations and accounting for their occurrences by way of tropological errors does not free even the wisest theorist of the linguistics of literariness from ideology.

I end this section with yet another question: Is Paul de Man's "The Resistance to Theory" itself a work of theory? The answer would seem to be "indubitably yes." If that is the case, however, then is "The Resistance to Theory," as de Man says all theory is, both a resistance to reading and a resistance to theory? I do not see how de Man can escape the ineluctable laws he lays down. What is sauce for the goose is sauce for the gander. When his seminar students asked de Man whether he escaped those laws, as I remember happening, he just smiled inscrutably. He did not say yes and he did not say no. What about my own essay, the one you are at this moment reading? I too must be caught in the same trap, but I leave it to my readers to spot where this happens in my discourse.

7. Reading Pictures in the Twilight of the Anthropocene Idols

> *Literature involves voiding, rather than the affirmation, of aesthetic categories. One of the consequences of this is that, whereas we have traditionally been accustomed to reading literature by analogy with the plastic arts and with music, we now have to recognize the necessity of a non-perceptual, linguistic moment in painting and music, and learn to* read *pictures rather than to* imagine *meaning.* (de Man's emphases)
>
> Paul de Man – "The Resistance to Theory"

When Paul de Man died in 1983, the Internet was just a futuristic dream. The digital revolution was just getting underway. Consequently, he had little or nothing to say about what would in just three decades after his death change almost everyone's life worldwide. Though de Man speaks eloquently in my epigraph for this section about *reading* pictures, he did not ever do that, either in seminars, as well as I can remember, or in his published work, whereas Derrida read a lot of pictures, for example in *The Truth in Painting*. De Man manages to talk brilliantly about Walter Benjamin's "Die Aufgabe des Übersetzers" ("The Task of the Translator") without mentioning Benjamin's interest in graphics, from *Ursprung des deutschen Trauerspiels* (*The Origin of German Tragic Drama*), through his use in the "Geschichtsphilosophische Thesen" ("Theses on the Philosophy of History") of Paul Klee's *Angelus Novus* as an emblem for the way we go forward into the future with our backs to it, facing the past as an ever-increasing pile of debris, and then on to his concern with graphics in the incomplete "Arcades" project. De Man's almost exclusive interest was in words on printed pages, though we know from a notorious reference in his Yale inaugural lecture that he much enjoyed the television show, *Archie Bunker*. He makes a vile pun and calls a deconstructionist an "archie debunker." His Yale audience groaned disapprovingly, as I well remember. They perhaps thought such frivolity was not becoming on such an august occasion. Many people feel threatened by wordplay, particularly ironic wordplay.

8. Is the Digital Revolution the Radix Malorum?

Whether de Man would have had things to say about the role of pictures in our digital age we shall never know, nor can I claim to fill that gap in this essay. His work was, as I have said, always surprising and unpredictable. Nevertheless, I turn now to the evident fact that all five of the features I mentioned initially as characterizing our present dire situation would have been impossible without the digital revolution, which I mention as the fifth characteristic of "these bad days."

Global climate change is not being caused by the digital revolution, but primarily by human-caused CO2 emissions, though these emissions are, as Tom Cohen persuasively argues, inextricably entwined with language systems that were necessary to the "anthropogenic" activity that brought about the CO2 emissions. This includes ideologically motivated refusal to recognize until it was too late that there might be anything dangerous about them. The evidence for climate change given by scientists depends on things you can do only on a computer or other digital devices, for example, measurement of CO2 levels, assembly of data about climates worldwide, statistical modeling of various scenarios for the future on the basis of that massive data, and so on. That data is too complex and too multitudinous to be manipulated with a slide-rule. If we still had only slide-rules we might be dumbly wondering why our storms are so much more frequent and severe, why fires have raged in Australia and in the United States south and west, why we have more and more violent tornadoes in the Mid West, and so on. Computer calculations give explanations and plausible predictions.

All the information about levels of unemployment and attempts to scare citizens about the deficit and to persuade them to give up Social Security and Medicare is based on manipulation of computerized statistics. We are deluged on all sides with figures that are the result of digitalization.

The global recession would have been impossible without the creation of subprime mortgages through computerized banking and investment transactions that then allowed the creation of "derivatives" and "credit default swaps." No more keeping track of transactions by local banks in handwritten ledgers. All transactions are digitized, for example those automatic computerized microsecond trades I mentioned in my opening section. When "the computers go down" at your local bank, all activity pauses.

Public opinion is now to a considerable degree manipulated by digital gadgets like television, Facebook, Twitter, and so on. I shall discuss below the way this breaks down the distinction between truth and lie by turning everything into a digitized spectacle.

Our wars are now fought to an increasing degree by drones, unmanned aircraft that are a triumph of digitization. Someone at a computer screen in Colorado controls the drone and aims it at the target, killing lots of people, including innocent women and children, as well as, in January 2015, Western hostages held by Al Qaeda in Pakistan. The digital revolution has transformed warfare. Our soldiers train as killers by playing video games.

Literary study is being rapidly revolutionized by digitization, as my use of online sources in this essay indicates. I rarely need a research library these days to do my work on the remote island in Maine where I now live most of the time. Many people who read literature at all these days read it in e-Texts on a computer or on a Kindle or on some other e-Reader. A printed book is part of the print epoch, the epoch of library collections and books on scholars' shelves. Paul de Man still lived all his life in that epoch. An e-Text, on the contrary, is part of the vast virtual reality we call cyberspace. This is a quite different mode of existence from that of a printed book.

Perpetually plugged in, dwelling in cyberspace: that is the way we live now. I am at this moment attached to my computer as a prosthetic device that makes it possible for me to write this essay and then to revise it more or less endlessly with the help of Word for Mac 2011, version 14.4.9.

Terms that have been used to describe the results of living in a more or less completely digitized world include "spectacle" and "simulacrum." An e-Text is a simulacrum of a printed book. A video war game is a simulacrum of a real war, but so are those repeated videos on television that illustrate the latest news about Mid-East violence with a shot of soldiers crouching behind a wall and firing automatic weapons. The same videos are used over and over, week after week. They are not films of something that the verbal report describes as happening just today or yesterday. They are, in any case, from the beginning simulacra, digitized videos. We are not witnessing a real war scene. Moreover, the film clips in question are shown repeatedly on NBC evening news interspersed with the ads I described in the first section of this essay. At least forty percent of NBC Evening News is made up of TV ads, most of them for a narrow range of drugs: Cialis, Restasis, Chantix, Crestor, Advil, Claritin, etc., each with its own barbarous name. These ads are shown over and over from night to night, in nightmarish repetitions. They show actors pretending to be doctors or happy users. The fictitiousness of the ads rubs off on the so-called news that is shown in brief bits as interruption of the real business of the thirty minutes, which is to sell Cialis, etc. The soldiers firing automatic weapons over a wall become by contamination one more depthless ad, in this case propaganda for the endless continuation of wars in the Middle East.

This meretricious depthlessness of television's virtual reality calls up from my mental databank three names: Guy Debord, Jean Baudrillard, and Maurice Blanchot. Debord of course for *The Society of the Spectacle* (1967, in French); Baudrillard for *Simulacra and Simulation* (1981 in French); Blanchot for his theory of the "image" in "Two Versions of the Imaginary" ("Les deux versions de l'imaginaire") and "The Song of the Sirens" ("Le chant des Sirènes"). The ideas of these three writers can hardly be perfectly reconciled, but they are to some degree in resonance. All were in one way or another influenced by Marx.

Both Debord and Baudrillard were in different ways sociologists who deplored what they saw as a bad new society dominated by advanced capitalism and by the images created through mass media, advertising, film, popular music, etc. Both Debord and Baudrillard were deeply influenced by the semiotics of their time. Both were photographers, Debord as a more or less professional film maker. He made in 1973 a film of *The Society of the Spectacle* (Debord 1973). For Debord "spectacle" is "a social relation among people that is mediated by images consisting of mass media, advertising, and popular culture" (Debord 2013).

For Baudrillard, present-day society today has reached his fourth stage in the development of simulacra. That stage is "pure simulation," an interplay of simulacra without any relation to "material reality" whatsoever. Stage four is a complete transformation of society by television, film, print, and now the Internet into depthless spectacle (Baudrillard

2013). Today we would add Netflix, Facebook, Twitter, and video games to this list.

Debord's "society of spectacle" and Baudrillard's society of pure simulation are strongly analogous, in spite of important differences in terminology and mode of argumentation that need to be kept in mind. Baudrillard was a professional philosopher, while Debord was a film-maker influenced by surrealism.

Blanchot's idea of the "image" is quite different from Debord's spectacle or from Baudrillard's simulacra. I adduce Blanchot partly to indicate that all twentieth-century theorists of the image were not singing the same tune. Blanchot presents a subtle theory of "the image" as the essence of the imaginary embodied in literary language. Speaking, for example, in a characteristic torrent of paradoxes, of Proust's breakthrough when two sensations coincided in a time out of time that made it possible for him to become a writer at last, Blanchot says: "Yes, at this time, everything becomes image, and the essence of the image is to be entirely outside, without intimacy, and yet more inaccessible and more mysterious than the innermost thought; without signification, but summoning the profundity of every possible meaning; unrevealed and yet manifest, having that presence-absence that constitutes the attraction and the fascination of the Sirens" (Blanchot 2003, 14; Blanchot 1959, 22). This formulation, in spite of important distinctions, is not entirely different from Debord's spectacle or Baudrillard's simulacra. Blanchot's imaginary is a dangerous vanishing point within which one might be swallowed up and disappear. This danger is figured in the threat to Ulysses, in Homer's *Odyssey,* of the Sirens' song. Blanchot tends to identify the imaginary with death or with an endless process of dying. The imaginary also exists as "the narrative" (*le récit*), as opposed to the evasions of the novel (*le roman*). Blanchot's examples in the essays I have cited are Ulysses in his approach toward, or refusal to approach toward, the real song behind the Sirens' infinitely luring song, Proust's Marcel in his search for (*recherche de)* lost time, and Ahab's pursuit of the white whale in *Moby Dick.*

9. Verbal as Against Visual?

Recent critical thinking has sometimes made a stark contrast between verbal media and visual ones. The contrast has often been couched in historical terms. What Debord and Baudrillard say is posited on a theory of history as a series of distinct epochs determined in a Marxist way by modes of production and distribution, and by the media dominant at a given time. This ideological presumption assumes that during the epoch when print literature dominated, that is, during the 18th, 19th, and part of the 20th centuries, the primary cultural media—books, magazines, etc.—were linguistic: printed words on paper pages. More and more as the 20th century progressed, however, this historicizing concept assumes, new primarily visual media gradually came into cultural dominance: photography, film, television, video, and now the Internet, with all its concomitants, such as video games. Even popular music online is often accompanied by videos, as though we can no longer listen without at the same time having something to look at. Before, we needed to be expert in reading printed words. Now we must be expert in deciphering the implications of visual images, such as those in films or video games. The need to be good readers of, say, printed literature has apparently lessened.

Few people would hold to quite stark a contrast, but most of us have some idea, however vaguely held, that verbal and visual media are quite different and require different academic disciplines ("film studies," say, as against "literary studies"), and quite different methodologies of interpretation. These seem to be quite plausible assumptions until you begin to think a little about the actual history of the two media.

Verbal and visual media have been from the beginning of Western culture mixed in various ways at various times. That is my primary presupposition in this part of my essay. Whatever there may have been originally of the ideographic in Western alphabetic languages has long since mostly disappeared, as opposed to an ideographic language like Chinese. The Chinese character for "exit" still looks, to me at least, like an open mouth or a doorway, though Chinese readers may well not notice the ideogram any more than we Westerners see the letter "o" as an open mouth uttering "Oh!" Mastery of a distinctive calligraphic style used to be a requirement in China for officials from the Emperor down to the lowliest bureaucrat. Recent scholars have, however, studied the way the visual aspect of purely verbal written or printed texts in Western languages changes to some degree the meaning we assign: type font and type size, roman as against italics or bold, capitals as against use of colored type, space between lines and in the margins, binding, marks of punctuation like dashes and exclamation points that are not voiced, and so on.

This was not quite what de Man meant, by the way, by his formulation about "the materiality of the signifier" (de Man 1986, 11). He meant, at least in part, the sound or appearance in any instantiation of a word like "day," the sheer fact that any word must have some form of material base. "Materiality" is a complex word in de Man's work. In any case, I am arguing for differences in meaning arising from the *way* words are materialized, that is, from the visual side of words on the page or on the computer screen. My computer has a long list of fonts that I can choose from in different point sizes. The creation of fonts is work for highly skilled visual artists. My composition of this essay on my computer involves a whole set of choices designed to make what I compose look to me right on the page. A newspaper headline is in large type so it looks as if it is shouting and is really important: ROMNEY CHOOSES RYAN AS RUNNING MATE. Nevertheless, even the most purely visual set of signs invites being read as if it were a kind of writing, as my citation from Paul de Man asserts: "... we now have to recognize the necessity of a non-perceptual, linguistic moment in painting and music, and learn to *read* pictures rather than to *imagine* meaning."

By imagining meaning, I imagine, de Man meant the transformation of the words on the page into mental visual images, as when we, in different ways for each reader, vividly imagine the faces, bodies, and surrounding scenes of characters in novels on the basis of verbal clues the novel gives. Examples are the reader's internal image of those waves crashing on the shore in the Interludes of Virginia Woolf's *The Waves* (Woolf 1963), or what Wallace Stevens' jar in Tennessee looked like in "Anecdote of the Jar": "And round it was, upon a hill.... [It was] tall and of a port in air" (Stevens 1990, 76). In spite of de Man's denigration of imagining meaning, I think what might be called a spontaneous internal cinema accompanies, for most people, the reading of a verbal text. But reading pictures also happens for those adept in purely visual sign systems. What de Man meant by reading, alert readers of de Man's "The Resistance to Theory" will know. He meant, as I have shown in the first part of this essay, attention to "the linguistics of literariness," that is, to the aberrational implicit assertions of figurative language. ("A ship is a plow.") A familiar example of using the linguistics of literariness to read pictures is the way we understand montage in film, the sequential juxtaposition of separate scenes, according to tropological rhetoric, as a metaphor, metonymy, or synecdoche. Eisenstein's crowd scenes in Eisenstein's *Battleship Potemkin* set side by side with a shot of swarming maggots is a famous case of this. Such *readable* tropes, however, are a staple of the purely visual side of films and other visual media, even when they are not accompanied by printed or spoken words.

10. Two Contemporary Examples of Pictures That Invite Reading

Here are two concrete examples of this *readable* visual imagery, one from a *New Yorker* cover, one from an advertisement in *Wired* (20.08). These pictures exemplify in obvious ways Guy Debord's society of spectacle and Jean Baudrillard's final stage of the world as interacting simulacra with no reality behind it or referred to by it. Both mix verbal and visual. The first picture, however, works as a critique of ideology, whereas the second uses our everyday ideological associations to sell a consumer product. Readable visual imagery can function both ways, as demasking aberrations and as reinforcing them for a particular purpose.

The recent brilliant *New Yorker* cover I have in mind (July 23, 2012) shows a nuclear family (father, mother, and two teen-age children, a boy and a girl) standing side by side in tropical clothes on a beach at the edge of the ocean, with a backdrop of palm trees. In the foreground is the shadow of someone taking a digital photograph of the group. He stands just where you are, dear viewer, which makes you the one taking the snapshot that forms the cover. All four of the family members are holding iPhones or some similar gadget. All are texting or tweeting or using Facebook. They are paying no attention whatsoever to the beautiful scene they have come to visit, nor to one another. So much for the salutary togetherness of the nuclear family and for the salutary function of vacations in the tropics! The facial expressions of the four persons, especially those of the father and mother, are tense, even anguished. They are grimacing or clenching their teeth with concentration. They are plugged in. They are mere attachments to their digital machines. That is the way we live now. Such people are perhaps what all human beings will be like in the twilight of the Anthropocene idols.

The readable meaning of this cover is attained without the use of a single word, except of course the words that tell the viewer this is a *New Yorker* cover of a certain date, with all that implies, and the signature of the artist in the lower right hand corner (Ulriksen). Moreover, as everyone knows, the smartphones the family holds are multimedia devices, holding verbal, visual, and audio files. "A picture is worth a thousand words." In this case the meaning is generated by the ironic and discordant juxtaposition of the beautiful tropical scene and the family all isolated from one another and all wholly absorbed in their iPhones. The viewer is put in complicity by way of the shadow in the foreground that shows the viewer's surrogate taking a photo with another smart phone. Ironic incongruous juxtaposition is a readable/unreadable trope. It is impossible to read it unequivocally. Irony is always to some degree unreadable, as de Man persuasively argues in "The Concept of Irony" (de Man 1996, 163-84). Such an ironic juxtaposition could also be expressed purely in words, or in a mixture of words and images.

My second example is from *Wired* (2008, August, 2012, no page number). The transformation of *Wired* has been truly amazing. It began as a McLuhan-like interpretation of media, featuring authors like Nicholas Negroponte of the MIT Media Lab, one of *Wired*'s founders (see 2013d). *Wired* now describes itself as providing "in-depth coverage of current and future trends in technology, and how they are shaping business, entertainment, communications, science" (*Wired* 2013, under "Subscribe," no longer given on the Website in just these words). A subscription to *Wired* gives you free access to all of *Wired* digitized on an e-Reader app for an additional $5.00 a year. No more talk about Creative Commons, open source, and sidestepping copyright limitations. *Wired* has become a spectacle, a panorama of simulacra itself in need of demystification, a capitalist tool. That transformation corresponds to the rapid expansion and commodification of the Internet, as it has turned into a full-blown Debordian spectacle or collection of Baudrillardian simulacra.

In the *Wired* ad I have chosen to "read" the meaning is generated by metonymies that turn from side by sideness to the similarity of metaphor. The owners of the registered trademark for *smartwater®* (*glacéau® smartwater®*), with its weird accent over the "e," have refused to let me reproduce the ad. They were perhaps right to do that, since my goal is to demystify the way the ad works, thereby to disempower it. Doing that is definitely not in their interest. I repeat the ® both to protect myself against failing to acknowledge that the trademark is registered, and as a mild irony. *Smartwater®* is, by the way, a division of *Coca-Cola®*.

Most people are so used to seeing and "reading" such ads, in magazines, on television, and on the Internet that they take their interpretation for granted. We just let such ads work their magic "unconsciously," which is what the ad-makers probably intend. As Marshall McLuhan long ago recognized in *The Guttenberg Galaxy* (McLuhan, 1962), *Understanding Media: The Extensions of Man* (McLuhan, 1964) and other books, however, "The medium is the Message" (or "mass age" or "mess age" or "massage," according to his puns) (Anonymous, 2013b; Anonymous, 2013c). McLuhan had little to learn from Debord or Baudrilard about mixed verbal and visual imagery. He "got it" already, with brilliant completeness, including the recognition that "spectacular" juxtapositions of visual images work like verbal tropes. In the case of my *glacéau smartwater®* ad, visual images and words combine to produce a complex meaning. The word "*smartwater®*" echoes the word "smartphone," in another covert allusion. Both, the words imply, have a species of superhuman intelligence or will make you smarter than you already are.

In the *Wired* ad, a beautiful and provocatively dressed woman with parted lips sits in an expensive leather-upholstered convertible holding a big bottle of *smartwater®* erect in her left hand. The traditional association in ads of sex with fast expensive cars is shamelessly exploited once more. The woman's relaxed right hand calls attention to her open, unbuttoned blouse, her almost exposed left breast, and to what is hidden by her clothes lower down, where her legs fork and where her dangling fingers point. She is looking just over the viewer's left shoulder, as if about to turn to look you straight in the eye. She has a tastefully elegant bracelet on her right wrist and a matching necklace. The model, Beth Rosenberg tells me, is the television and film star Jennifer Aniston (Anonymous,

2013a). The caption at the bottom reads, with tasteful alliterative terseness, "good taste travels well."

The message is clear. You would be smart to drink *smartwater®* because it will give you good taste in two senses, the taste of the water (which has electrolytes, and does taste good) and the social good taste that will earn you a ride with the beautiful woman, also in two senses. A promise that drinking *smartwater®* will enhance your manliness may also be read in that proffered bottle. It must work sort of like Cialis. I speak from the perspective of a heterosexual male, as you will notice. I suppose a good percentage of *Wired*'s readers are in that category. Many names on the masthead are female, however. Stories about women software entrepreneurs and other such professions are included. Female readers, straight or lesbian, would no doubt respond in a way different from me to the ad's allure.

All these double meanings are achieved by means of visual or verbal puns, similes, metonymies that become metaphors: the taste of the water is like the promised taste of the attractive woman, sex is like a fast car, and so on. To carry this interpretation even a step further, you might read the snap-buttoned leather box (the enclosure for the convertible's roof) in the foreground as an invitation to a further act of unbuttoning, or of entering an enclosure.

In this ad everything is turned into image and becomes a siren song promising, in Blanchot's words, "movement toward a point—one that is not only unknown, ignored, and foreign, but such that it seems, before and outside of this movement, to have no kind of reality" (Blanchot 2003, 7, trans. slightly altered; Blanchot 1959, 13). I really know that drinking *smartwater®* will not lead to possession of the sexy woman in her expensive car, but the ad nevertheless generates a movement in me toward the unfulfillable promise these images make. I know I am just looking at the photograph of Jennifer Aniston, who probably survives on carrot-juice and who takes whatever pose the photographer requests. The whole thing is a sham, a simulacrum, a spectacle, a "model," as in "fashion model," "late model car," model train," and "model pupil" (that is, a paradigm for others). My deconstructive knowledge, nevertheless, achieved by my knowledge of the linguistics of literariness, as de Man asserted and as I too believe, does not by any means prevent me from responding to the ad's fraudulent appeal.

The ad is made more powerful and persuasive, by the way, by being placed side by side in *Wired* with a great many other such ads for high-end commodities of all kinds. Their allure rubs off on one another, in another distributed metonymy. My *smartwater*® ad, for example, is followed, after a second masthead page, by an ad for the *asics*® *Gel-Lyte33*™ running shoe, that by an ad for the Hyundai 333-hp GDI V6, that by an ad for Fidelty Investments, and so on throughout the issue, which is about half made up of advertisements. This spectacular transformation of digital media into ideological "performatives" that persuade you to act in some way or other, for example to buy lots of *smartwater*®, is especially evident in the use of television, Facebook, Twitter, etc. in political campaigns, as in postings and television coverage already ubiquitous for the Presidential race of 2016. The terms "truth" and "lie," as Debord and Baudrillard in different ways argued and demonstrated, have ceased to have relevance to this complex tissue of mixed media assertions. If you say often enough that Barack Obama is a Kenyan socialist Muslim bent on destroying capitalism, a lot of people will believe it, just as many people will associate Mitt Romney primarily with that vacation drive to Canada with his dog strapped to the roof of his car or with his spectacular media gaffes. The dog on roof event apparently did happen in material reality, but it becomes a mediatic simulacrum when used over and over in campaign rhetoric.

11. Mixed Media Forever

I return now to my claim that multimedia in different forms and mixes have characterized verbal texts from the beginning. That means it is a mistake, in my view, to think of a radical change in the twentieth century from print media to graphic media, from printed novels, say, to films, with each requiring different disciplines and methodologies of interpretation. What is needed rather is a flexible procedure of "reading" appropriate to each given mixture of media at a given historical epoch. Paul de Man, in the essay I cited at the beginning of this essay and have then read in detail, called this procedure, as we know, a use of "the linguistics of literariness," that is, the rhetoric of tropes, to read all sorts of cultural phenomena. If he is right about the need to learn to *read* pictures, this claim would apply to television ads sponsored by conservatives claiming that lowering taxes on the rich will create a lot of jobs: the long-discredited "trickle down" theory. Changes in media have been all along rather a matter of complex changes in balance through the centuries among forms of expression that always have been, and always will be, mixed.

Most scholars will agree with this if they think a little about it, but the recent supposed shift from print to graphic has been a powerful ideological presupposition. From the mixture of verbal and graphic in carvings on Greek and Roman tombstones and funerary monuments, to medieval illustrated manuscripts, to the sumptuous graphic title pages and other illustrations in early printed books (modern title pages still commonly have graphic elements), to seventeenth-century emblem books, to Hogarth's great eighteenth-century mixed media graphic works (so brilliantly interpreted by Ronald Paulson [Paulson, 1971]), to printed novels that from the beginning have often, though not always, had illustrations, for example the great illustrations for Dickens's novels by Cruikshank and Phiz, to Dante Gabriel Rossetti's mixed media compositions juxtaposing poems and pictures, such as his *Aspecta Medusa*, to the sinister illustrations by Aubrey Beardsley for Pope's *The Rape of the Lock* or Wilde's *Salomé*, to the inclusion of photographs in some early 20th-century multi-volume "sets" of English writers like Thomas Hardy and Henry James, to the multitudinous illustrations in the magnificent Cook and Wedderburn edition of Ruskin's works in thirty-nine volumes in 1903-12, to the combinations of print novels and films of those novels or television versions that have characterized later twentieth-century forms of mixed media, to the wonderful efflorescence of comic strips such as *Peanuts* or *Pogo*, to graphic novels based on the conventions of comic books, such as Art Spiegelman's *Maus* or the Japanese "manga" graphic novels, many "translated" into English, though that is hardly the word for changing a graphic Japanese word for "pow!" or "bang!" into its English equivalent, since the original depends so much on the way the Japanese characters look when inscribed on the graphic page. (I have decided it would be pedantic to list in my Works Cited all the items mentioned in this paragraph. You can reach any one of them through Google or Wikipedia.)

The most spectacular (in both the idiomatic and Debordian senses) recent example of mixed media is, of course, the Internet, or Cyberspace generally. Cyberspace mixes verbal, visual, and auditory materials in a hyperbolic way. Almost all the Websites I list in this essay mix verbal and graphic materials, and many have auditory components too. Adepts at navigating cyberspace (or falling into it) must have skills in "reading" this latest example of mixed media.

If I want nowadays to teach or write about Imre Kertész's *Fatelessness* (Kertész, 2004) or Ian McEwan's *Atonement* (McEwan, 2001), or even novels by Austen, Dickens, Hardy, James, Conrad, and others, I more or less am obliged to consider the films and BBC television versions of all these. I say "obliged" and "ought" because in my experience more and more students and faculty around the world will have seen the film but will not have read the book. If you want their attention and understanding in a lecture or essay you had better say something about the film.

I think part of the reason for the sharp separation between verbal and graphic in some people's minds still today derives from the way publishers during a relatively brief period of high modernism tended to publish novels without pictures, for example the novels of Conrad in the Dent Edition, or the Hogarth Press edition of Virginia Woolf, or the paperback editions of novels by Faulkner (though those had lurid covers). Virginia Woolf's *The Waves,* for example, in the Hogarth Press edition, has no illustrations beyond the elegant cover designed by Vanessa Bell (Woolf, 1963). On the other hand, it is available for free in a searchable e-Text that turns it to some degree into a graphic presentation, a display on the computer screen (Woolf, 1931). Older novels, for example those by Dickens, Trollope, or Hardy were usually published during the high modernist period without their original illustrations. That led teachers and students (me for example) to forget or ignore the role those illustrations originally played in the generation of meaning.

In doing this I was forgetting the children's books that did so much to inform my early sensibility, my pleasure in puns, wordplay in general, and my spontaneous ability to "read" pictures. Such books would include especially, for me, Lewis Carroll's, *Alice in Wonderland* and *Through the Looking-Glass* (Carroll, 1865; Carroll, 1871), Kenneth Grahame's *The Wind in the Willows* (Grahame, 1908, read by me as a child in a later edition with the Ernest H. Shepard illustrations of 1931), A. A. Milne's Pooh books, also with Shepard illustrations (Milne, 1926; Milne, 1928). I remember as much the glorious Shepard pictures of Humpty Dumpty, or of the tea party in which the dormouse is dipped in tea, in the *Alice* books, or Rat "messing about in boats," or Toad tearing down the road in an open car calling for "Speed! More Speed!" in *The Wind in the Willows,* or Piglet, in one of the Pooh books, terrified by Pooh fallen in a pit with his head stuck in a honey jar who therefore looks to Piglet like a "horrible heffalump" or a "heffable horrilump." But I must desist.

Henry James, in *A Small Boy and Others*, puts his finger on the truth about these mixed media creations as read by young people. Speaking of the Cruikshank illustrations for *Oliver Twist*, James, with wonderful astuteness, asserts: "It perhaps even seemed to me more Cruikshank's than Dickens's; it was a thing of such vividly terrible images, and all marked with that peculiarity of Cruikshank that the offered flowers or goodnesses, the scenes and figures intended to comfort and cheer, present themselves under his hand as but more subtly sinister or more suggestively queer, than the frank badnesses and horrors" (James 1913, 120). Cruikshank's Sikes, James is saying, looks, paradoxically, wholesome and sane compared to his Mr. Brownlow or his Oliver.

12. Just How Has the Internet Transformed Literary Studies?

Today print culture is fast fading. It is being rapidly replaced by digital culture of all sorts. Most people in the United States do not any longer read Shakespeare or Dickens or Emily Dickinson unless forced to do so in school. They watch Fox News on cable television, or the PBS evening news, or BBC sitcoms, or, a few people, television adaptations of classic English novels. Today people play collective or single-player video games online. They spend hours every day using cell phones, iPods, iPhones, iPads, email, Facebook, Twitter, and wireless laptop computers. They use smileys or emojis and they text, or indulge in "txtng," even while driving a car, to their great peril. They communicate by Skype with a simulacrum of a friend on the screen.

Features of the so-called post-modernist sensibility, specialists in it like Fredric Jameson say, are subjective depthlessness, absence of unified selfhood, lack of affect, and the loss of any historical sense. These are among the results of our society of spectacle. One must be careful, however, not to fall into some naïve version of irresistible technological determinism. The new gadgets can be used in all sorts of ways. They limit but do not absolutely determine the use that is made of them.

A concomitant of the digital revolution, which has taken only twenty years to happen, has been great confusion and uncertainty in the humanities. What should we humanists study and teach? If we teach anything like the old curriculum, we are teaching things that have less and less relevance to the actual lives of our students. In Victorian England (my special field), middle-class ideology, beliefs about gender roles, courtship and marriage, class divisions, and so on, were both transmitted and to a considerable degree created by novels, by works written by Jane Austin, George Eliot, Charles Dickens, Elizabeth Gaskell, Anthony Trollope, and a host of other popular novelists. Now such beliefs are passed on and to some degree created by films and by radio and television, including talk shows on TV and radio. The latter have such increasing power in politics that some people have said that the actual heads of the Republican Party in the United States are people like the hosts on Fox News, that is, those who exploit the media. Cyberspace is full of the most amazing vitriolic lies.

Most people these days have not read Jane Austen at all. They know her work, if at all, only through the latest BBC adaptation. An online ad from the University of Minnesota Press listing new books for 2010 in "Literary and Cultural Studies" did not have one single book on literature. Book publishers, including university presses, appear, on the good evidence of dwindling sales, to have concluded that little or no market exists for old-fashioned literary studies, whereas a book on the cultural history of Botany Bay (one book on the Minnesota list in 2010) is assumed to have a market. Now (April 2015) the University of Minnesota Press no longer even has a catalogue for "Literary and Cultural Studies." That catalogue has become just "Cultural Studies 2015."

One result of these amazingly rapid changes, which I call "prestidigitalization," has been the overnight change in the humanities from literary study to cultural studies. My decision in this essay to use the linguistics of literariness to read two images from popular culture is an example of this. Younger scholar-teachers, especially, want to study and teach what really matters to them and to their fellow citizens. This is all to the good. Cultural studies, however, has not yet quite become a well-organized discipline, as once was, for example, the study of British literature or of medieval European history or of German literature. Cultural studies straddle the humanities and the social sciences, as do sociology, anthropology, and "media studies." No widespread agreement exists about just what is the best training, the best curriculum, to prepare scholars to do "cultural studies." By contrast, I once could have told you exactly what courses you need to take to prepare yourself for a career as a specialist in Renaissance literature or in Victorian literature. We knew, or we thought we knew, what you needed to know.

Some amazingly good things, however, have been brought to literary studies by the digital revolution. The ease of writing and revising on a computer is one of them. This essay is an example of that. Moreover, it is no longer necessary to be at a major university with a big library to do serious research and criticism. This is a powerful form of democratization. Much literature is available online, if you happen to want to read it that way, for example work by all of the writers I have mentioned in this essay, as well as large amounts of secondary literature about literature: all of the novels by Dickens, Henry James, Anthony Trollope, etc. etc. I worked not long ago on Franz Kafka and Imre Kertész. More or less all of Kafka's work in German is available online in searchable form. When I wanted to check the equivalent Hungarian word for "naturally," and its frequency in the original, as against its recurrence in the English translation of Kertész's wonderful Holocaust novel, *Fatelessness,* I found a searchable version in a minute by Googling for it. That location has since mysteriously disappeared, by the way. That is probably because after Kertész received the Nobel Prize in 2002 the printed version in Hungarian became valuable again, too valuable to give away free on-line as an e-Text. I ordered the film version of *Fatelessness* from Amazon and had it in my computer's DVD slot in a few days. Recently I could not, to my shame, remember the location in Shakespeare of a phrase I recalled, "the beast with two backs." In two minutes I found it by way of an online concordance. It is Iago speaking in *Othello,* act one, scene 1, line 126.

The Internet, the World Wide Web, and Wikipedia are fantastic resources for even the most traditional forms of humanistic scholarship. I use Wikipedia a lot. As with any encyclopedia, one has to be skeptical and check it against other sources, but I have found Wikipedia to be amazingly accurate when I have checked its entries against other sources. One "spectacular" result is that I rarely have to borrow a book from the quite comprehensive University of California library system. I can do the scholarly and critical work I want to do just about as well in Sedgwick, Maine, or on Deer Isle, Maine, as I could in Irvine, New Haven, or Baltimore. I have emphasized this by giving all the URLs for the many online sources I have consulted to write this essay.

Emailing, though I spend an inordinate amount of time doing it, has changed my life. It has put me in more or less instant contact with scholars and students all over the world. Through email I have created my own *virtual* communities. I stress "*virtual.*" Essays and book manuscripts are sent to me by email attachment, though I still have difficulty reading books and essays on a computer screen or on my Kindle for Mac. That difficulty is diminishing.

I am certain that using email, the Internet, and the computer has changed my personality, as well as the way I write essays and books. I have not, however, found it all that easy to be precise in identifying just what the difference is between doing literary study exclusively with printed books and essays and doing it with the help of the Internet. I think Debord and Baudrillard may help understand that difference. If I read a printed book that act ties me to the print epoch. It puts me back into a time when, however problematically, verbal and graphic creations were assumed to be in one way or another representational. They were taken as linked to the real extra-verbal and extra-graphic world by some form of mimesis. When a given printed text, Woolf's *The Waves* for example, is made available online in an e-Text, it is subtly transformed into image. It is transmogrified willy-nilly into one simulacrum among the billions of simulacra floating around in cyberspace in our society of spectacle. That is a huge difference. Whether I can work with such materials and remain true to the McLuhanesque commitments I share with the founders and early editors of *Wired,* or whether I will inevitably become subdued to what I work in, like the dyer's hand, as rapidly as were those in charge of

Wired as it became more and more commercialized, is another question. Attention to the linguistics of literariness may help me keep my head. In any case, working with digitized materials as opposed to print materials means submission to a quite different technological or spectacular regime, even though both print and the results of prestidigitalization are different forms of mixed media.

13. Imagine Paul de Man Online in the Twilight of the Anthropocene Idols

Whether Paul de Man, if he had lived another thirty years, would have turned to "reading" the Internet and other forms of digitization is an impossible question to answer. It is a fruitless speculation. He was pretty firmly committed to study of "the linguistics of literariness" as it is present in printed texts. He belonged heart and soul to the print epoch. He was a printed-book man. Trying to think of de Man before the screen of an Apple laptop, with his fingers on the keyboard, as Derrida is shown in films we have of him—it is impossible. De Man would never have fallen for that new technology. That technology is neatly symbolized by the Apple logo: an apple with a bite taken out of it, as in the Fall of Adam and Eve. Charlie Gere has noticed the implications of this (Gere 2013). De Man never even got as far using a typewriter, though I don't remember him ever expressing the active hostility to it that Heidegger does in his Parmenides lecture (Heidegger 1942-43). Jacques Derrida has written eloquently about Heidegger's preference for handwriting in "Le main de Heidegger" ("Heidegger's Hand") (Derrida 1987; Derrida 2008). De Man wrote all his essays out "in long hand," as almost all scholars from Plato through Rousseau to Walter Benjamin and Heidegger did. Nietzsche, on the contrary, bought in 1882 and used at least briefly a typewriter, a Malling-Hansen Writing Ball. As confirmation that I am well and truly fallen into cyberspace, I googled "Nietzsche, typewriter" and "Heidegger, typewriter." In each case I found in a few seconds a substantial list of online texts on the topic, as well as (surprise!) ads for typewriters (Nietzsche, typewriter, 2013; Heidegger, typewriter, 2013).

De Man had no experience with cyberspace. Nevertheless, his proleptic insight into the nature of the ideological mystifications that now surround us on all sides, mostly in digitized form and usually with a visual component, can give those who have outlived him and who have taken a bite out of that apple a path to follow. We can attempt to unmask today's ideologies and account for their occurrence, as I have tried to do with the ad from *Wired*. As part of this unmasking project we need to learn how to *read* pictures, as de Man said.

De Man does not, however, promise that this unmasking and accounting for will cure us of the ideological aberrations that are rapidly turning the twilight of the Anthropocene into total darkness. Far from it. An example of that failure to cure is the resistance to Paul Krugman's Op-Ed pieces in the New York Times. These argue authoritatively, with empirical evidence from what has happened in Ireland, Great Britain, Spain, and Greece, that austerity measures implemented in the United States, as the Republicans are determined to do, would likely bring on another recession, raise unemployment, and cause widespread social misery. States like Kansas and Louisiana that have tried austerity have more or less gone bankrupt. A good percentage of the blogs in response to Krugman's essays, however, just go on saying in different forms that slashing Medicare, Medicaid, and Social Security, while lowering taxes on the rich and deregulating banks and investment firms will bring good jobs and prosperity to all. Another example is the large number of people who buy the National Rifle Association's absurd solution to gun violence (such as the Newtown massacre of twenty children and six adults, and many other such events since Newtown): arm everybody, including especially elementary school teachers, with military style assault weapons equipped with large ammunition magazines. Denying that climate change is happening and denying that it is "anthropogenic" is only one more of these delusional mystifications, though perhaps the most catastrophic.

Just unmasking these delusions will not end them. De Man does nevertheless offer ways of understanding what is happening to us, as year-round heat waves and violent winter storms increase, as the Arctic, Antarctic, and Greenland ice melts, and as the water rises up to our necks, as it did during Hurricane Sandy for those on the Jersey Shore and in Lower Manhattan.

14. Anachronistic Reading

In this final section of my part of this book, I present a reading of a poem by Wallace Stevens (Miller 2010). Although I cannot hope to match the rigor and inventiveness of de Man's work, I present it as exemplification of my own way of doing readings of literature in the light of de Man's work. Since this essay was first published,[37] as my contribution to an issue of *Derrida Today* in my honor, a lot of water has gone over the dam, but the changes have been mostly for the worse. The Affordable Healthcare Act has given millions more Americans affordable healthcare, though the Republicans have done everything they can to repeal it. They will go on doing so, while universal single payer healthcare, such as all other first world countries have in some form, remains the proper and obvious course. Our two houses of Congress are more or less dysfunctional, with constant threats of government shutdown made by the Tea Party crazies in the House of Representatives. An immense number of American citizens get their political opinions from conservative (and blithely lying) radio and television talk shows and television news. The gap in wealth between the top one percent and the rest of us continues to increase to unheard-of proportions. The Supreme Court has made a number of disastrous decisions, for example "Citizens United," which allows the very rich to buy our politicians with unlimited secret contributions. The candidates vying for the Republican presidential nomination for the 2016 election are, every one of them, disasters. Any one of them, if elected, would do immense damage to American democracy, for example by abolishing or privatizing Social Security and Medicare if they could do it, by cutting taxes on the rich and on corporations, by deregulating banks and other financial institutions, and by doing nothing to mitigate climate change, which many of them continue, in defiance of all the evidence, to deny. Climate change, meanwhile, is happening with astonishing rapidity, with wild fires, floods, droughts, water shortages, and the like happening with increasing frequency and intensity in the United States, not to speak of the melting of the Arctic ice-cap and the Antarctic glaciers. The role of the humanities in higher education continues to diminish. President Obama has just announced (October 16, 2015) that we are not going to pull all of our troops out of Afghanistan after all. Altogether, we are in an increasingly parlous condition as a nation, on the way to becoming an oligarchy dominated by a few very rich people and families, as well as engaged in perpetual wars. This final section is my small contribution, to quote what I said at the end of the previous section, to procedures using literary works as a means of "understanding what is happening to us, as year-round heat waves and violent winter storms increase, as the Arctic, Antarctic, and Greenland ice melts, and as the water rises up to our necks." My focus is primarily on climate change, as is consonant with this book as a whole.

Literary works program or encode their future readings, though in an unpredictable way. This exemplifies Michelet's dictum that "Each epoch dreams the one to follow," or Percy Bysshe Shelley's claim in "A Defence of Poetry," that poets are "the mirrors of the gigantic shadows which futurity casts upon the present." The poet, says Shelley, "not only beholds intensely the present as it is, and discovers those laws according to which present things ought to be ordered, but he beholds the future in the present, and his thoughts are the germs of the flower and the fruit of the latest time."[38] A free reading would try to identify the way a poem or other literary work mirrors the future. Such a reading, it follows, is anachronistic. It takes possession of the old work for present uses and in a new context.

A poem encrypts, though not predictably, the effects it may have when at some future moment, in another context, it happens to be read and inscribed in a new situation, in "an interpretation that transforms the very thing it interprets," as Jacques Derrida puts it in *Spectres de Marx*. The effects of a transformative reading exceed any intentions or intended meanings that may have been in the poet's mind when he put the words down on paper, or, in Wallace Stevens's case, dictated the poem to his secretary. A poem is only completed when it is freely read at some point in the future. It takes two to tango, the writer and the reader. Each reading, moreover, is unique, even if the same reader makes several readings of the same work at different times. The reading creates the meaning retroactively, by a species of metalepsis or what Freud called *Nachträglichkeit*, cause after effect, the cart before the horse. For Freud the initial traumatic event only becomes traumatic when its effect is triggered by a much later event that recalls, echoes, or repeats the first event. Another way to put this is to say that the poem, though not the poet, foretells, foreshadows, foresees, prefigures, or even brings about performatively the meaning and force it comes to have. Franz Kafka feared his stories and novels might bring about the catastrophe of the Holocaust they obscurely foresee. Therefore he could not finish his novels and wanted all his manuscripts burned, in an attempt, vain as it turned out, to abolish their performative force.

Just what would an anachronic, ahistorical, anachronistic, inaugural, proleptic, transormative reading, such as I am demanding, look like? I take Stevens's "The Man on the Dump," from *Parts of a World* (1942), as my example. The poem, John Irwin reminds me, may echo the film of 1936 starring William Powell and Carole Lombard. William Powell plays the man on the dump who is hired as butler by a rich socialite, played by Carole Lombard. Here is a link to the poem http://www.poetryfoundation.org/archive/poem.html?id=172209

Well, what can I say about this poem? What does the poem say?

On the one hand, the poem tells a plausible realistic story. The speaker, whoever that is, let us call him the Hartford poet and insurance executive Wallace Stevens, "now, in the time of spring," has gone to the dump at sundown and moonrise, perhaps the municipal dump of his home-town. He observes the everyday objects of American consumer life that have been discarded in the dump and carries on the endless meditation that was Stevens's interior life and the concomitant of his poetry. That meditation, as transferred to language or generated by it, alternates, as was usual with Stevens, between an ornate richness of metaphorical transfiguration and a minimalist shedding that attempts to discard or eject all images, to "wipe away moonlight like mud," as he says in another poem. The speaker beats an old tin can, a lard pail, no doubt found on the dump, as a rhythmical accompaniment to his meditations.

On the other hand, no thoughtful reader can doubt that "The Man on the Dump" is more than the realistic report of a visit to the Hartford dump by Wallace Stevens in 1942. The verbalized meditation is the point or substance of the poem. The poem's material base must never be forgotten, nevertheless, since it provides the "material" that Stevens's language transfigures. That transfiguration takes two chief forms in the poem.

One is the characteristically gaudy tropological transformations. The moon is seen as a woman named Blanche who places the sun in the sky like a bouquet of flowers, or, more precisely, like a "corbeil," name for "a sculptured basket of flowers or fruits used as an architectural ornament" (American Heritage Dictionary). The poem uses a wild assortment of tropes, a big bouquet of them: metaphors, similes, metonymies, synecdoches, prosopopoeias, comparisons of more and less, alliterations that call attention to the material base of language ("bubbling of bassoons"), synesthesias, prolepses, analepses, metalepses, puns.

"The Man on the Dump," moreover, turns back on itself to reflect on this process of verbal transmogrification and, ultimately, to reject it—almost. Of "the freshness of morning, the blowing of day," the poet asserts, "one says/That it puffs *as* [a simile; my italics] Cornelius Nepos reads, it puffs/More than, less than or it puffs like this or that." The word "puff" apparently refers to the intermittent breezes of a spring morning, but that puffing can be said to be more or less or like all kinds of things, this or that, even like the way Cornelius Nepos reads. Nepos was a Roman biographer, who lived from ca. 100-24 B.C. His only surviving work is the *Excellentium Imperatorum Vitae* (*Lives of the Eminent Commanders*). That work is certainly "puffs" in the sense that we speak of outrageously hyperbolic praise of a newly-published book as a "puff." If the fresh breezes of a spring morning can be said to be like Cornelius Nepos's puffs of these murderous military men, then almost anything can be said to be like almost anything else.

The other form of transfiguration is to see the junk in the dump as figures or what Stevens calls "images" and what Marx, and I, would call the residue of consumer fetishism's ideologies. The waste items in the dump, so carefully itemized by Stevens, are not exactly symbols. They are synecdochic examples: newspapers used as wrappings for bouquets, the wrapper on the can of pears, the cat in the paper bag, the corset, the tea box from Esthonia, spring flowers that must be discarded when they wilt (azaleas, trilliums, myrtle, viburnums, daffodils, blue phlox), old automobile tires, lard pails, bottles, pots, shoes, and grass. This motley assortment is archeological signs of the social life of those people who have used and have discarded them, just as is the case with Roman or Mayan dumps archeologists disinter and analyze.

Stevens's term for the way these things in the dump are ideological signs of commodity fetishism, and of the state of technology at that time, is, as I have said, "images": "The dump is full/Of images." The movement of "The Man on the Dump" is a complex temporal to and fro or give and take that attempts to move toward a repudiation or annihilation of these images by the man on the dump. That rejection leaves him "The Latest Freed Man," as the title of the second poem after this one in the *Collected Poems* puts it.

In "The Man on the Dump" this movement of imaginative "decreation," that sees things not as images but as they are, takes place quite suddenly, just after the gorgeous description of the way men and women copy dew in their clothes and jewelry: "One grows to hate these things except on the dump." The poem opposes two kinds of time, the perpetual rhythm of sunset, moonrise, and then sunrise, in an endlessly repetitive present ("The freshness of night has been fresh a long time. The freshness of morning ..."), to the time of culture in which things are manufactured, festishized as social values, and then thrown away, to remain as archeological or historical testimony in the form of images on the dump. We come to hate these images, to find them disgusting, like so much garbage. Walter Benjamin's Angel of History, in the "Theses on the Philosophy of History," faces towards the ever-growing debris of history at his feet as he is propelled backwards into a future he cannot see. Stevens's man on the dump has his back toward history and has repudiated its images in "disgust" in order to face toward a future that will, he hopes, be purified of ideological mystifications. He knows, however, that always more such images will come and will need in their turn to be repudiated and junked, put on the dump, time after time, in an endless task of purification.

The result of rejecting the trash is that "everything is shed." In that moment of minimalist decreation so important in Stevens's poetry, one sees (or thinks one sees) things as they are, without trope or deviation, in a double purification of both seer and seen: "the moon comes up as the moon/(All its images are in the dump) and you see/As a man (not like the image of a man)./You see the moon rise in the empty sky."

Such seeing without ideological distortion is extremely difficult and precarious. The last stanza, the most difficult lines in the poem, attests to that. After the lines about beating an old tin can, lard pail, as a way of getting at "that which one believes," which is "what one wants to get near," the poem ends with a strange series of questions. It also returns to the gaudiness and tropological extravagance of the opening lines. Are these just rhetorical questions or are they real questions? It is not easy to tell, though the question is important.

If the stanza returns to the extravagant language of the poem's opening, after the bareness of a disgusted rejection of all the trash, it returns with a difference. To get near to what one believes is to get near to the self. The poetic self is, for Stevens, the home of the imagination. That imagination creates what Stevens calls the new "supreme fictions" by which, according to him, we all ought to live. For Stevens, as for Percy Bysshe Shelley, poets are the legislators of all mankind, or at least the makers of new ideological images, as well as being those who put in question the old images.

For Stevens these supreme fictions are, for United Statesians, specifically American fictions based on our own landscape, our own birds and beasts. Stevens celebrates, in "Sunday Morning," his version of the ideology of the American Adam making a home in the wilderness: "Deer walk upon our mountains, and the quail/Whistle about us their spontaneous cries:/Sweet berries ripen in the wilderness" (Stevens 2006, 70). Stevens knew as well as did Louis Althusser (if not Marx, who was beguiled by the hope of an end to ideology), that when one ideology or set of cultural images is rejected in disgust and sent to the dump, the resulting bareness lasts only a brief instant. It is then replaced by a new set of cultural images. Stevens wanted the poet, that is, Wallace Stevens himself, the man on the dump, to be the begetter of this new ideology.

The final stanza of "The Man on the Dump" obscurely projects as a hypothetical event that performative action in its series of puzzling questions. I take it the ear is superior to a crow's voice in the same way that the imaginative self is superior to anything it can sense. That is because the ear can transform, through words, those sounds, for example the ear's hearing of the cawing of crows, into articulate speech. That speech says anything the imagination likes, in sovereign domination. The next question I take to be an allusion and repudiation of Keats's nightingale. We have no nightingales in New England. That means all those meanings Keats ascribes to the nightingale are alien to the New England mind, however much they inhabit it, as in the echoes of Keats in this poem. Since the nightingale was alien to us, it tortured the ear, packed the heart, and scratched the mind. These are strong images of violence done to body, feelings, and mind by the foreign bird. Here and now, in Hartford in 1942, the ear solaces itself in the sounds made by peevish birds like the grackle. The grackle is our equivalent of what the nightingale was for Keats. The poet says one finds peace on the dump, in a philosopher's honeymoon. This is perhaps because the epistemological quest so central to philosophizing is appeased at last. Peace comes when all the images are on the dump, in a general purification of the mind. Then you can see "as a man (not like the image of a man)." That brief moment is like a honeymoon because a marriage of imagination and reality is possible in that instant. Such a marriage is Stevens's perpetual goal. That goal is reached only at brief moments if at all.

Stevens's philosopher appears covertly later in the stanza. In the third stage in the cycle from infatuation with images to throwing all those images on the dump to the creation of a new mythology, the poet murmurs "*aptest eve*" as he sits among the mattresses of the dead and the other junk on the dump. An evening free of images is aptest for the new performative action of poetic language. The poet says "*invisible priest*" when hearing "the blatter of grackles" because the poetic imagination is an invisible ubiquitous priest. That priest takes on the role of transubstantiation once played by priests in the literal sense. An example is his transformation of the blatter of grackles into poetry. That transmutation is happening in the lines we are at that moment reading. Such efficacious priests as we have these days are the poets. In an appropriation of sacerdotal language, the poet is, like James Joyce's Stephen Daedalus, a "priest of eternal imagination, transmuting the daily bread of experience into the radiant body of everliving life" (Joyce 1964, 221). The phrase "invisible priest," to my ear, is another echo of Keats, in this case the sonnet, "Bright Star". That poem contains the marvelous figure of "nature's patient sleepless eremite,/The moving waters at their priestlike task/Of pure ablution round earth's human shores" (Keats n.p.).

That leaves the puzzling phrase "*stanza my stone.*" Why does the poet cry these words when he has ejected all the false images, pulled the day to pieces in a rejection of "*the* truth," to replace that false hope with visions of a world that is always in parts? I claim that the reference is to the philospher's stone, sought throughout the middle ages and Renaissance. It was a magic stone, "as hard as stone and malleable as wax," that would turn base metals into gold. "It was also sometimes believed to be an elixir of life, useful for rejuvenation and possibly for achieving immortality." In Latin, "philosopher's stone" is *lapis philosophorem*, in Greek *chrysopoeia*.[39] "*Chrysopoeia*"! That means, literally, "making gold," but "poeia" is of course the root of our words "poetry" and "poet." Stevens's cry, "*stanza my stone*," is a triumphant claim, just before that last line rejecting "the the," that the poet's stanza, his words on the page, is the philosopher's stone that turns the base metal, for example the blatter of our New England grackles, into the gold of the supreme fiction embodied in the words on the page. This new fiction will, in a powerful speech act, replace all those outworn images on the dump. Poetry is the philosopher's stone. We need no other.

~

I have sketched out a reading, a commentary or exegesis, of "The Man on the Dump." You could call it a hermeneutico-poetic reading. It is not yet, however, the *free* reading, that performative reading, reading as work, as *oeuvre* that does something with words, for which Jacques Derrida calls.

How many people really care, nowadays, for the subtleties of a poem by Wallace Stevens written in Hartford, Connecticut, in 1942, and readable in the context of social and cultural history of that time and in the context of Stevens's other work? I mean, really, how many people truly care about such antiquarian excavations, such digging in the dump of history? I want to know why I should read this poem today, now, in October 2015, and what use I can make of the poem for decisive action now.

I turn now, briefly, to a more "free" reading, a performative interpretation that sees the old text as somehow a proleptic foretelling of a present situation. For Paul de Man a poetic reading is always allegorical in the specific sense de Man gave to the word "allegory," as opposed to "symbol." "Symbols" are hermeneutic in the sense that their meaning is their spatial reference to something non-verbal, as we might take Stevens's poem to refer to the Hartford municipal dump in 1942. An allegorical sign, for de Man, however, in the famous formulation in "The Rhetoric of Temporality," is part of a temporalized structure in which the sign refers to another earlier sign of which it is a repetition, "in the Kierkegaardian sense of the term." In place of de Manian allegory I put a proleptic reading that sees a text as prefiguring a future event that comes to seem what the text predicted, foresaw, or forecast. This future chiming would also be a sign to sign relation, an anticipatory allegory or, perhaps, a prophecy or, perhaps, a miniature apocalypse in the etymological sense of an enigmatic unveiling of what has not yet happened. Poetics has no good term for this different form of sign to sign relation across a temporal gap in the future rather than the past. "Prefiguration" is perhaps the best word we have, with an echo of Biblical *figura,* the Old Testament prefiguring the New. Such a miniature apocalypse is an unveiling that hides as much as it reveals. Models for this kind of reading might be my commentary, in another essay, following other critics, on the last chapter of Kafka's *Der Verschollene, The Man Who Disappeared.* That incomplete novel is commonly called, without Kafka's authority, *Amerika.* I see that chapter as an ironic foreshadowing of the *Judentransporte* that took so many unwitting millions to the death camps.[40] Other examples would be seeing, as Russell Samolsky does, Conrad's *Heart of Darkness* as a foreshadowing of genocide in Rwanda (with the crucial difference that the Rwanda massacres were not the direct consequence of colonial violence, as were the massacres in *Heart of Darkness*). Samolsky also sees Kafka's "In the Penal Colony" and Coetzee's *Waiting for the Barbarians* as foretellings or premonitions not just of torture under apartheid in South Africa but of United States's shameful, abominable torture of prisoners in Abu Graib and Guantánamo Bay. We should read each in terms of the other. One could not easily read Coetzee's novel today without thinking of our use of waterboarding and other forms of torture. A final example of such prospective allegory is Derrida's "interpretation that transforms the very thing it interprets" of Marx in *Spectres de Marx.* Derrida appropriates Marx for our present global situation. He dreams of a New International that would move toward the ever-receding horizon of "the democracy to come." What Derrida says goes well beyond anything Marx said and even disagrees with it on a crucial point. He rejects Marx's ontologizing of material means of production as irresistibly causing ideological mystifications.

What in our present situation does "The Man on the Dump" presage or foretell? It is impossible to read Stevens's poem thoughtfully today without seeing how Stevens's dump with its single human presence anticipates our present condition, all six and a half billion and counting of us. We are collectively living on a planet that is becoming one gigantic garbage dump. Just as Kafka's single protagonist, Karl Rossman, foreshadows the six million victims of the Holocaust, so we are all now, all the time, men and women on a universal dump. Stevens lived in that happy time before we became aware of climate change, global warming, widespread species extinction, including the possible self-extinction of *homo sapiens*, glacial melting, rising sea-levels that will before long flood my two acres of beautiful shore property on Deer Isle, Maine, not to speak of much of Florida and Manhattan, increasingly violent storms—hurricanes, tornadoes, typhoons, floods, forest fires—all caused by our collective human dumping of waste. Bats and frogs are already becoming extinct. Polar bears are endangered by the melting of the ice floes on which their survival depends. Scientists predict that by the end of our century over half the world's species will have vanished in the sixth great extinction. This one will, however, be caused not by asteroid-impact, like the one that triggered the fifth great extinction at the end of the Cretaceous period, but by clever human beings, *homo sapiens*. Human beings, those wise creatures, as they have become more numerous and have colonized the earth, have brought devastation everywhere. This devastation has come along with the benefits of civilization, such as frozen processed foods, full of toxic chemicals, and broadband Internet service (Kolbert 2009, 52-63).

Stevens's dump was relatively sequestered in one place. It was composed of relatively benign and biodegradable materials: mattresses, corsets, flowers, newspapers, bottles, shoes, grass, pots, tin cans. Our present world-wide dump is made not only of all the non-biodegradable plastic and mercury-filled computer junk in our local dumps proper, but of pesticides and chemical fertilizers that poison our agricultural soil and make it incapable of absorbing carbon-dioxide, of all the carbon dioxide in the atmosphere from automobiles, coal-fired electricity plants, and other sources that is a chief contributor to global warming, of all that methane from domestic cows and from landfills, of all the smoke from forest-clearing, of all the nuclear waste that the clever scientists do not seem to have foreseen would be almost impossible to dispose of safely. Huge rafts of floating garbage, much of it plastic, litter our oceans and wash up on our beaches. The oceans' PH level has risen, killing fish, coral, and seaweed worldwide. We have already taken for food about half the edible marine creatures in the whole ocean. Melting permafrost and melting glaciers reduce the reflective surface of the earth and accelerate global warming. Hardly a week goes by without some new evidence that global warming is already real and already probably irreversible. Another big chunk of the Greenland icecap falls into the sea, or another evidence of rising temperatures is recorded.

All this devastation is striking confirmation of Jacques Derrida's law of auto-co-immunity. This law is his claim that human cultures have an unconquerable penchant for self-destruction through their effort to make themselves safe. This, says Derrida, is analogous to the way the human body, in auto-immune diseases, destroys its own tissues. The body does this by way of an immune system that is supposed to protect the body from alien invaders, viruses and the like. All we human inhabitants of the earth are busy big time destroying the earth as a viable habitat. Practically nothing is being done to stop climate change in any way at all commensurate with its rapidity. This is so in spite of the development of windpower, solar power, geothermal power, electric automobiles, organic farming, and the like. The latest Clean Energy Bill before the United States Congress has been viscerated by the lobbyists for coal-fired electricity-producing plants and their ilk. Legislators worldwide continue to dither and delay. Deniers of global warming still abound. Scientists, meanwhile, keep saying, in surprise, that climate change is happening even faster than they had predicted.

Just as Stevens figured by way of the junk on the dump the way the mind of the man on the dump was contaminated by cultural "images," so the mind of almost everyone today is infected by vignettes about global warming and about the way the world has become one huge garbage dump. A recent pamphlet I found in my local doctor's office is called "Stop Trashing the Climate." It has a photograph on the cover of a bulldozer on an apparently limitless garbage dump. The text gives statistics about the damage to the environment and the increase in global warming caused by our habits of dumping garbage and the use, for example, of incinerators or methane collecting apparatuses rather than extensive composting.[41]

I watched in the '80s and '90s from my house in Irvine, California, the building up of a gigantic landfill about a mile away in a ravine across a small valley. Lines of trucks came to dump there day after day more trash, much of it, apparently, vegetation debris, tree branches and the like, but other parts apparently loads of ordinary garbage. Eventually a methane collection system was installed on the top of the hill, across the freeway. A housing development has now been built on the dump site. I am glad I do not live there. The inhabitants are, literally, men and women on the dump.

I think also about that old story of the garbage barge in the sea off New York and New Jersey that kept going from one place to another trying, unsuccessfully, to get some garbage dump to take it, like Kafka's Hunter Gracchus on his wandering death barge.

I think of the decades spent, so far unsuccessfully, to get agreement about how and whether to put nuclear waste deep in Yucca Mountain in Nevada, or of that serio-comic story of the shipping container full of plastic ducks for childrens' bathtubs that fell off a ship into the Pacific Ocean. Plastic ducks kept coming ashore in distant places for years. Who knows how many are still floating around out there, like miniature Flying Dutchmen.

I think of my own local dump on Deer Isle that is now divided into piles of mattresses, piles of metal and plastic junk, recycling containers, kitchen garbage containers, and piles of wood and cardboard that are periodically burned, putting yet more carbon dioxide and toxic chemicals into the atmosphere. It will not escape my readers that the photographs I have taken of these sights make me the man on the dump, with camera.[42]

I think also of that poisonous lake in Tennessee whose dam burst recently, covering a large area with toxic chemicals and heavy metals.

I think of that picture of a beach on one of the Hawaiian Islands, covered with junk from the open ocean. As Tom Cohen observes, "what may now be the largest man-made feature detectable from space" is "the vast floating island of plastic debris that spans a large part of the Pacific ocean."[43]

I think of those recent stories about the breaking off of a piece of the Wilkins Ice Shelf in Antarctica, a piece as big as the state of Connecticut in the United States.

I think of that recent Pixar/Disney animated film, *Wall-E,* in which the earth has become uninhabitable, a huge garbage dump. All the remaining human beings have fled earth seeking some new habitable planet, leaving behind a solitary robot, Wall-E. Wall-E was programmed to pick up garbage and compact it into meter-sized cubes, which he goes on and on doing, mechanically. He builds great walls and buildings of cubes made of compacted garbage.

That the trashed earth, in this animated film, is left only with a robot doing what it has been programmed to do might be taken as a compact emblem of the way climate change, though initiated by men and women, is a mechanical process. That process is now proceeding apace beyond our control, just as human auto-immunity is a non-human process. It is a result of something that has gone wrong with an immune system. That system works automatically, for our good or ill. Our consciousness does not have any control over it. The immune system just goes on working, blindly, as chemicals and genes program it to do. Another analogy is way computer nerds concocted the computer programs that were necessary to create complex derivatives and credit default swaps. Now, however, these programs operate, as bankers now probably truthfully say, beyond human comprehension or control, like robots gone wild. The banks and other financial institutions claim, perhaps truthfully, that they have no way of knowing how many billions or even trillions they are collectively in debt beyond their available assets. The parallel with climate change is betrayed in the way we speak of "toxic assets" and "global financial meltdown."

Many human beings live today not just *on* the global dump, but prosthetically connected to it much of the time, as when we watch TV, use an iPod or an iPhone, or a computer connected the Internet. While giving this paper in oral form, I used an Internet "connection" by way of the Wild Blue Satellite service to bring down from cyberspace the photographs that are signs we are trashing the climate. The global trash heap is us. We are not separate from it. It is an extension of our bodies and minds. Stevens's protagonist is a single man meditating on a local dump. We are now six and a half billion men and women plugged into a worldwide toxic garbage dump. Our future, perhaps, if *Wall-E* is right, is to vanish, leaving behind a single surviving garbage-compacting robot on a global trash heap.

Another version of our modern man on the dump is that *New Yorker* cover by Dan Clowes for the June 8 & 15, 2009 issue. It is entitled "Future Generations." The picture shows a little green man from outer space. He has come to earth from his flying saucer overhead. The background shows an uninhabited Manhattan overrun with vegetation and green vines or green slime. This alien is sitting in a dump entirely made of smashed computer parts, CDs, and the like, and surrounded by broken walls. He is reading a book.

The importance of the model of the machinal or mechanical as opposed to the organic or consciousness-organized is hinted at by the importance these days of the word "apparatus." The word names a machinelike assemblage put together by human beings to do some kind of work or other. We speak of a "photographic apparatus," or of a "dredging apparatus." Wall-E is a robotic apparatus for compacting garbage. We also, however, speak of our "security apparatus" as a name for the interlocking agencies that led to torturing at Abu Graib and Guantánamo Bay. Michel Foucault used the word *dispositif*, translated into English as "apparatus," as a name for the machine-like working of the whole social-legal-governmental-financial bureaucratic assemblage in a given society at a given time. A willing worker for such an apparatus is called, in Sovietese, an "apparatchik." This apparatus, said Foucault, is "a thoroughly heterogeneous set consisting of discourses, institutions, architectural forms, regulatory decisions, laws, administrative measures, scientific statements, philosophical, moral, and philanthropic propositions—in short, the said as much as the unsaid. Such are the elements of the apparatus."[44] Today Foucault might have added to this litany of the wielders of power the media, including the Internet. Althusser includes the media among ISAs, Ideological State Apparatuses, in his famous essay on ideology's interpellations. It is a feature of this all-embracing social apparatus that it operates on its own, robot-like. What it brings about is "Nobody's Fault," to cite the title Dickens first presciently intended for *Little Dorrit*, with its Kafka-like presentation of the Circumlocution Office. The latter is the model of an efficient bureaucracy that just goes on and on, like the Energizer bunny, doing its intended work of procrastination, in a way parallel to the legal system in Kafka's *Der Prozeß* (*The Trial*). As Barack Obama is discovering, it is extremely difficult to alter such an apparatus in any fundamental way. It is a Herculean job, for example, to clean the Augean stables of our present catastrophic health-care apparatus and replace it with the only sensible form, single-payer universal health-care run by the government and paid for by taxes. Obama fell into the trap, during a press conference about health care, of answering a question about Henry Lewis Gates Jr.'s arrest. Obama's response (saying the police acted "stupidly" by arresting a man in his own home who had committed no crime beyond protesting to the police officer) gave the media a chance to create a great storm of controversy. This successfully turned attention away from the urgent necessity of health care reform if the United States is not to go bankrupt. Suicidal auto-immune responses like this can be seen every day in the media and in the behavior of our legislators.

I think, finally, of that recent story in the Huffington Post about an expedition to the summit of Mount Everest confirming that the Himalaya glaciers are rapidly melting, that the glacial lakes there are getting bigger and threatening to burst their ice dams and inundate the villages below. The melting ice and snow have exposed all the debris from previous expeditions. "Mountaineers in the past habitually left behind their climbing gear, tents, empty oxygen bottles and food wraps, leading to Everest being dubbed the 'world's highest garbage dump.'" The expedition brought down to base camp from near the summit 11,000 pounds of garbage.[45]

The last thing in Stevens's mind when he wrote "The Man on the Dump" in 1942 was the coming of catastrophic climate change as a result of the whole earth becoming a garbage dump. The poem itself, however, I claim, "dreamed the epoch to come," or, rather, a free or performative interpretation today would see Stevens's poem as foreshadowing our situation by that reverse allegory I have described. In such an allegory present image metaleptically echoes beforehand future event, preposterously, in an act of foreseeing. Climate change in itself is invisible. It is only visible in the signs of it. Those signs, such as the verbal vignettes I have presented, and indeed the host of photographic images to which they may be linked, resonate allegorically, in a reverse allegorical sign to sign relation, with the verbal signs in Stevens's poem. One way to define my reading would be to call it an allegorical interpretation not of Stevens's poem, but of all those present-day images of global climate change as interpreted prophetically by Stevens's poem.

What then of the move in the last stanza of Stevens's poem to a claim that the poet is the supreme artificer who will create the supreme fictions by which we can live better lives? The good poet constantly makes up new image clusters that will just as constantly find their way to the dump. I do not think that heroic concept of the poet's role has much purchase these days. A free or transformative reading of Stevens's poem, such I have sketched out, would rather claim that a reading for today is either an instigation to urgent action to do what we can to "stop trashing the climate," or a way of facing up to what is happening to us, as the water rises to inundate our coasts and as species after species goes extinct. You can take your pick. You can be either a watcher or a doer. I claim this essay is a form of doing things with words and other images, even if what it does is only to cry, "Get ready. The end of the world is coming." It is, perhaps, distinctively human to bear witness, to use language to testify that catastrophic climate change is happening and is going to happen.

Notes

Preface

1. Here, again, we would extend a point made by Latour. The correct response to climate-change denial is not to insist upon absolute certainty and the cold hard facts of science, for it is the notion of a truth that cannot be contested or narrated otherwise—the criteria of being beyond *all doubt*—that places an undue and impossible pressure on science. What we are contesting in today's narrative of the Anthropocene is that genesis is anthropogenic; what has taken place, over centuries is a series of thefts, substitutions, justifications, reparations and triumphs. If today we accuse "humanity" or "anthropos" of scarring the planet, and then attribute this intention to an agent—capitalism—such that "we" might then be otherwise, we are taking part in a history of inscription and rhetoric.
2. Or, a snowstorm is a snowstorm: it does not mean that global warming is a lie, nor is it a sign that we must act now to survive. Snow storms, unemployment figures, stories about illegal immigrants stealing our jobs, statistics regarding the dependence of the US economy on illegal immigrant labor, the shooting of unarmed African Americans by law enforcement officers, the election of the first black US president, the IPCC report on rising sea levels, alongside more dire warnings—all confront us, and need to be read. These are neither signs of a conspiracy, nor of an absolute imperative for humanity to regroup; not one of these inscriptions is the cause, or ground, or truth of the other.
3. What we contest is that one matter stands for or speaks the truth about another.

Chapter One

4. Norman Myers and Andrew H. Knoll, writing in 2001, cast the question of mass extinctions in the context of a *future* biotic crisis—that the "anthropocene's" drastic deletion of biodiversities reduces the combinant store of options for post-anthropoid "life," and that even the wash and rinse period of five million years or so when mass extinctions begin sorting out new arcades of life forms would be crippled: "The biotic crisis overtaking our

planet is likely to precipitate a major extinction of species. That much is well known. Not so well known but probably more significant in the long term is that the crisis will surely disrupt and deplete certain basic processes of evolution, with consequences likely to persist for millions of years. Distinctive features of future evolution could include a homogenization of biotas, a proliferation of opportunistic species, a pest-and-weed ecology, an outburst of speciation among taxa that prosper in human-dominated ecosystems, a decline of biodisparity, an end to the speciation of large vertebrates, the depletion of "evolutionary powerhouses" in the tropics, and unpredictable emergent novelties. Despite this likelihood, we have only a rudimentary understanding of how we are altering the evolutionary future. As a result of our ignorance, conservation policies fail to reflect long-term evolutionary aspects of biodiversity loss. Human activities have brought the Earth to the brink of biotic crisis. Many biologists (e.g., refs. 1–5) consider that coming decades will see the loss of large numbers of species. Fewer scientists— witness the lack of professional papers addressing the issue— appear to have recognized that, in the longer term, these extinctions will alter not only biological diversity but also the evolutionary processes by which diversity is generated. Thus, current and predicted environmental perturbations form a double-edged sword that will slice into both the legacy and future of evolution" (Meyers and Knoll 2001, 5389-5390).

5. At issue is, of course, not just the flat, morose, aura-less *words* that have circulated in this zone, architected for failure, nor the mediacratic as well as psycho-telic streams of climate denialism—basic modes of affect manipulation, perpetual ambiguation, invitations to side with a crafted "we" and defer incomprehension, theological invocations, and the parade of American double-talk that have earned the leader of the "free world" such skepticism and cynical resentment abroad as the "one superpower." Its lead Senator on committee an oil shill decrying perpetual "climate hoax," its Florida governor forbidding the use of "climate change" for state employees, even a Yale program dedicated to climate communication settling for a poll on which is better "climate change" or "global warming" for the commoner (the latter, because it's hot at least). But, then, we were addressing in parallel the academic, humanist, Leftist, and critical theory blinds that, broadly and with exception, maintained or do the same denialist posture due to investment and, of course, recoil. As Dipesh Chakrabarty remarks in closing out a "post-colonialist" modem before the logics of extinction and climate change (still, *human on human*, utopist, empathetic to its imagined "others"), these twentieth-century modes and historicist modes not only fuel the ecocidal acceleration but relapse before its thought (Chakrabarty 2009). Most interesting, since the recoil against so called linguistic pre-occupations and "deconstruction" specifically, Derrida's choice to leave this unaddressed in his late and last writing, for reasons one might speculate as both structural and risky to his project as a legacy work, mimed unaware climate denialism—and,

inversely, damaged his currency by avoiding just this intersection. Is it not surprising, by now, that we have no *rhetoric of "climate change"* written yet?

6. In choosing the proper name *Anthropocene* over, say, *Sinocene*, when the latter arguably has the longest running terrestrial civilization but is the most numerous and consequent to twenty-first century impasses, a subtle adjustment is made, a card held under the pack, beyond the obvious Western nominal land grab—though the Chinese might know *not* to take ownership. That adjustment is the assignation of this "epoch" to an alphabeticist writing trajectory, a telos-inflected grammar (sentence narration), a nihilist-monotheist arc, a technological "era of the Book" and theologization of a priest-hermeneut class, and the abstractions of an "Enlightenment" premise never lived and posited for "them" (as Conrad's Kurtz might explain). Ecocide would—misleadingly or not—not be assigned to the temporalities and techno-organizations of the pictographic cognitive script, with its re-aligning dashes and disjunct vocals, absent psychotic mono-gods, overrun by "industrial revolutions" and the fertilities of intra-state competition and dueling commerical imperiums.

7. "If... collective individuation, like psychic individuation, is a process of adoption, and though this adoption is in part linked to what Freud called 'identification', *the question of the we is that of its consistence insofar as it does not exist* (just as spirits do not exist, any more than phantoms or ghosts, which does not prevent them from being effective: this is what is referred to as symbolic efficacy)" (Stiegler 2014, 12).

8. Someday, when the prospect of an angel doesn't get readers hot and bothered to identify with it at any price, Benjamin's chastened scrutinizers will realize they'd been had, and that the too-renowned Angel of History Benjamin gives a sentence or so to before literally annihilating it in his *Theses* was a front and bait for the very identificatory and recuperative takes on him he disowned in the first theses—that, say, of the Marxist dialectician (Adorno) and the theo-hebraic allegorist (Scholem). Benjamin had already identified Klee's "new angel" with something more ferocious, without anthropomorphic accoutrement (no ruddy cheeks, nice wings, no fake sorrowful back glances at his own hypocrisy toward the dead readers looking to his face). In the essay on "Karl Kraus" it appears, succinctly, as the *Unmensch*, and it is particularly angled against Nietzsche's Overman, whose rhetorical fireworks Benjamin regarded still as romantic hedging—human, all too human, Zarathustra's animals notwithstanding: "This must be borne in mind if one is to understand the urgency with which he decried the dancing pose affected by Nietzsche—not to mention the wrath with which the monster (*Unmensch*) was bound to great the Superman (*Ubermensch*)" [Benjamin 1986, 452]; "One must have followed Loos... , heard the stellar Esperanto of Scheerbart's creations, or seen Klee's New Angel (who preferred to free men by taking from them, rather than make them happy by giving to them) to understand a humanity

that proves itself by destruction" [456]; "Neither purity nor sacrifice mastered the demon; but where origin and destruction come together, his reign is over. Like a creature sprung from the child and the cannibal, his conqueror stands before him not a new man—a monster, a new angel.... Angelus—that is the messenger in the old engravings." [457] In turn, in the rhetorically bifurcated *Thesis IX*, this *Unmensch* returns not as the duped and annihilated "Angel of History" (or, more in synch, historicism), but what displaced him and his face—what is called, in climactic jargon, the vortex of a "Storm" (*Sturm* replaces the grammatical subject in the successive sentence). At once destroying force without any possible anthropomorphism, tied to technic accelerations ("Progress") and to inscriptions ("the old engravings").

9. I had been asked sometimes to clarify an argument put forth that Paul de Man is best read as a writer *avant la lettre* of the era of climate change—though I concede that I resent a bit being prodded to this question. I have changed my sense of this, as well. At least, I no longer *recommend* pursuing what I use de Man to index as a missing piece in the thinking of "climate change," at least, not without goggles, a semio-biotic lens, and a sense of the comic void of any recuperative reflex. In a sense, I *resign* from this assignment, but I will try to indicate why. If I no longer recommend de Man as a hidden counter-stroke, it is not because the argument is not cogent. Rather, it is because of the way that it is cogent. To give this a little piquancy: can one posit something like a *literary structure to "climate change,"* one that even guarantees ecocide, due to a regime of hermeneutics instituted reactively to the hyposcripts out of which perceptual grids, referential conventions, and agreed upon memes form a discrete but totalizing closure—for which the critical probes of extending itself to new others remains a strategy of sustainability?

10. However, part of the *passive* violence of the digital transcription of "everything" (archives, "data") is the quiet retirement of the *humanualist* template (clearly run down by the time Derrida stalked its mechanisms), that is, of the hand in favor of fingers, digitalia. (This may supplant the *humanualist* with, in Jonty Tiplady's phrase to me in an e-mail, the "who-digitalian.") One types, no longer handling a pen or stylus to scratch. Tone no longer holds (conceptually). This loss of the privilege of hands would echo a retirement of an economy of face. Fingers make contact point (typing), fan into spider dances, reduce the affective counters and expanse of the skin-sack to pixels and points—redesigned as the "mimetic" screen. See Tom Cohen, "Tactless—the Severed Hand of JD" [Cohen 2009].

11. This does no more than "see" *light* as a technic (pyrotechnic) itself, as what precedes the trope of a *fiat lux*, and the binary division of a light defined by its binarization through darkness, blackness, an absence that confirms its presence: but the technics of "light" includes, nuclear fission aside, waves, frequencies, spectrums of invisibility, hence interval, absence, "blackness" in advance—for which, in turn, the tropologies of blackness are inadequate (and

privative). For discussions assuming, advancing, or interrogating this premise as that of digitalization, see Cubitt, Palmer and Tkacz (2015).

12. "Corporations are, theologically speaking, institutions of death. They commodify everything—the natural world, human beings—that they exploit until exhaustion or collapse. They know no limits. There are no impediments now to corporations. None. And what they want is for us to give up. They want us to become passive. They want us to become tacitly complicit in our own destruction" (Hedges 2013).

13. https://vidrebel.wordpress.com/2011/11/10/catherine-austin-fitts-the-black-budget-and-the-leveraged-buyout-of-the-world-using-stolen-money/

14. The "epistemology of tropes"—as de Man characterizes it—is given textbook rendition in parts of Nietzsche's fragment, and the premise for de Man's use of the text (a fragment of a sentence) can be read as its displacement. For de Man, this critique is essential, unavoidable to arrest cognitive automatisms and highlight the essentially subjectless position of mnemonic epistemologies, but not an end in itself. Its trajectory leads beyond itself, as "tropes" are put in question collectively as substitutive chains that, in the end, merge, mime one another, find themselves totalized (in a vortex), or as neutered as the term *ideology* today (which these parallel as *aesthetic* confabulations). They are also diversions and evasions of "material" events or inscriptions that, outside of phenomenality, project and give rise to the then antithetical play of tropes (*doxa*, "perceptual" consciousness, ideology, cognitive automatism, hermeneutic *relapse*). That tropes guide or foreclose affect and cognitive effect is hardly news to tele-marketing whizzes (consumerist and political), whose science of designer metaphors and corn-brewed "affect" is trumped only by post-communist nations' unrivalled psy-ops and propaganda science (it helps if one's legacy is that there is *only* propaganda—that is, rhetorics of power—but, over time, doesn't help if that's reduced to crude slogans and mono-cultural fantasy). Enter de Man, who puts his foot in this door but feigns timidity over what lay on the other side, something which would not accommodate dialectic binaries, personification or perceptibility. While the most transparent title in this regard is "The Epistemology of Metaphor" [de Man 1979], this first premise—that substitutive tropes and their death-cycles program cognitive and affective processes—permeates *Allegories of Reading* [de Man, 1979]. The "epistemology of tropes" precedes so-called ideology critique, since like the former, "ideologies" replace one another as content for a similar platform or regime of reference, perceptual hallucination, narrative editing. One can criticize one from the position of another, seemingly, but not exit the vortex: which explains why both the utopic left and the utopic right (Banksters and their minions) fuel the ecocidal acceleration. Here is where the late de Man both extends and muddles what is no longer a dialectic puzzle or temporal construct when, in "Kant and Schiller," he improvises (it is a transcribed

talk) that, despite the "irreversibility" of what appears a movement from an essential "critique of the epistemology of tropes," one does not arrive at a "performative" alternative (that would simply occur regardless) nor, he insists, does one revert ("reversal") into the tropological foreclusure—what facilitates, say, climate change denialism and invisibility. The hermeneutic fold back that is more or less reflexive or inescapable is only a "relapse," not a reversion, something is clawed forward, and he will leave this movement hanging in the figure of the "passage" between an epistemology of tropes and the language of power, a "passage" that is in perpetual question except as passage. My guess is that the word "relapse" had a special resonance for the cancer-stricken de Man, whose green light prognosis at the time would very soon, and suddenly, convert to kill him. So there is the implied door-stop: this does not all disappear with the "relapse," the hermeneutic recuperation (as occurs routinely in Derridean "deconstruction"): something happens; one, we, it, does not revert, it remains, a la Benjamin, a one-way street, and so on. Thus de Man in "Kant and Schiller": "the passage from a conception of language as a [cognitive tropological] system... to *another* conception of language" [de Man, 1997, 132]. And: "There is a difference, a progress made, it cannot return despite this refold—the "relapse" harbors the seeds of a next fumigation: "[A]nd as such [it] is not a reversal, it's a relapse. And a relapse ... is not the same; it *has to be* distinguished in a way which I am only indicating here but which would require much more refined formulation—the recuperation, the relapse *has to be* distinguished from a reversal"[de Man, 1997, 133]. And yet, in fact, it serves to extend the domain of the reversal, if you like, giving inverse birth, which is to say marking, to the *irreversible*. The improvization needs a supplement, certainly from the twenty-first century and following the pan-archival digital transcription and totalized "data" harvesting since accelerated, as if the latter had read de Man on "aesthetic ideology" for pointers, inversely, on the capture of "inscriptions" themselves, the control of the sensorium as of assigned debt. There never was nor needed to be "reversal," the "relapse" is all there ever was and simulates reversion; in fact, the "irreversible" requires the step beyond that had from the earliest inhabited de Man's artifice (his dissertation, "The Rhetoric of Temporality" and before). But if the "epistemological critique of tropes" is inescapable, it also expires as a sort of painted doorway. All tropes appear substitutive maneuvers with interchangeable ends, a perpetual system one can neither exit from nor exceed by outwitting. This parallels so-called "ideology critique": one can "criticize" one ideology from the perspective of another, which would replace it, but both are modeled on, and substitute different "contents" for, the same aesthetic regime and programs of reference, identification, promissory time, and the same conversion of inscriptions into hallucinatory totalities and perceptual projections. Thus the premise of *aesthetic ideology*, which departs from a shared program of reference and perception out of which so-called ideologies occur (interchangeably, like tropes). This is why the progressive leftist and the petro-banker converge in fueling the ecocidal acceleration, without exit. De Man, Claire Colebrook has opines, did not go far enough—and he did not, in any case, attend a twenty-first century environment where pan-archival digital transcription and pre-emptive streaming would both enforce and disassociate closed circuits of "tropes" as a controlled perma-environment, a mode of capture still refining itself and anticipating the leap to simple implants. It would not be that, say, heliotropism pervades and misinforms the figural chains of the West, as a revelation to consider, but that the illusion of a trope of "light" retro-posited by that regime yields, today, an entire template of "Enlightenment" stitchings detached, fully, from the mafiatized institutions that re-run those rhetorics as amoral gameboards—moreso, as the flood of climate refugees, random "genocides," the writing off of peripheral "states" and populations, and so on, proceeds into an era of managed extinction (for the disposable populations as the bottle neck approaches. De Man's disinterest in psycho-analysis was the same as that in the ping-pong of "ideology critique," each never left the epistemology of tropes 2.0.

15. One can decide, as Bernard Stiegler does (2013a), that this coalesces in the *arche*-cinematic event of the cave paintings, a cinematized model subject to torch-waved shadows, the inception of a vision tracking movement (and identifying mimetic animemes with the hunt (and its hermeneutics), the inception of extinction events (the megafauna of the first screens would be the first to go, regurgitated, perhaps, in Spielberg's dinosaur *re*-animation trope for cinema), duly codified in Plato's allegory of the cave and resonant in the cavernous multiplexes of today, the cemetery parks of cinema—finally dissolved into the direct neural engagement of interactive screens, digital implants, honed mediacratic trances and the industrial redesign ("proletarianization") of the *senses* themselves.

16. Any *telepolis* simulates and swarms with algorithms of identification and coded referents, traversed by the strictly formal calculus of "data": everything is of the order of inscription so long as it appears anything but that, offered as the efficiency of rhymed anaesthetizing drug hits. Moreover, since this final avatar of the *arche*-cinematic cave is totalized, it not only censors the rushes and practices pre-emptive forensics, soon to be genetic, but both scripts and hacks memory regimes and "reality." As we see today, *telepoloi* may be encompassing but arrive in competing variants, edited or warring with one another ostensibly (since that may always be for marketing purposes). The *telepolis* precedes nationalist "content" (cultural and linguistic), we see divergent warring regimes and models positioning for the resource wars gathering. China would re-cycle a sort of "cultural revolution" lite *meme*, eschewing textbooks with "western values" together with corporate presence—and as the "western" *telepolis* is de-centered or exposed as running on Potemkin "Enlightenment" fronts ignored the corporate imperatives and the cartels of a klepto-mediacratic elite, others re-center: as, with a hyperbolic and derisive edge, Putin's "Russia" has, closing its media and, increasingly,

internet off to the despised "west," generating a war economy for a return that returns to the tsars and Stalin, to World War II and the medieval princes (anything, that is, except the Mongol occupation). The phrase "media ecologies" had been used, perhaps too tamely, to suggest what have become small and large, differentiating and warring, closed media "economies" that are, simultaneously, in a permanent war setting. The *telepolis* unadventurously names the simulant place without place of the shaped and enclosed digital streams, the gaming of totalized tropes dissociated from inscriptions and a real that is fully hacked—then taken for the latter (Cohen 2014).

17. "Autobiography as De-Facement" (de Man 1979b) is de Man's titular or iconic treatment of de-facement, but its implied logics permeates unrelated formulations, such as, in "Genesis and Geneaology (Nietzsche)": "The tradition is caught in a non-dialectical notion of a subject-object dichotomy, revealing a more or less deliberate avoidance of the moment of negation that coincides, for Hegel, with the emergence of the true Subject" (de Man 1979, 80).

18. "[T]he time of hyperobjects is an age of hypocrisy, weakness, and lameness"; and: "Humans have entered an era of hypocrisy" (Morton 2013, 24). Morton's para-ethical shift to grand style hyperbolism is, in fact, warranted by the semiosis claimed for the *hyperobject*—in effect, Morton's friendly death hug to *OOO* by taking the "object" hyper (the subject-object rhetoric as perma-trap). The underlying import of the claimed disconnect between the current sensorium regimes and their referential skeins points, however, in the alternate direction from the viscously ensconcing *hyperobjects* beyond phenomenal access: that is, toward modes of hyposcript, such as *inscriptions*, that are sealed off from perceptual access—where Morton's examplary hyperobject, *oil*, passes into the liquidation of carbon and imprints (ink), and the permeability of organisms and forms of animation in this perpetual projection loop. If the era of hypocrisy and weakness accords, say, with Wall St. insider casino plays, the end-running or gaming of all "laws," or the "short-circuiting" (Stiegler) of pure technics taking itself as target, we have returned to the pre-moment of initiation of the *arche*-cinematic, "light," and the archival premises of "anthropos" in a modum of pure, faux Promethean *theft*.

19. Bernard Stiegler's reclamation of an *arche-cinematics* (as "older" than Derridean *arche*-writing), and its claim to span up to the "death" of cinema in its digital transcription in today's multiplexes opens, decisively, a non "Western" technics that is only partly exploited as Stiegler invests in the socio-tropological imaginaries architected from Simondon's (faintly yet in fact non-Nietzschean) weaves of "trans-individuation" (Stiegler 2010, 2014). It may be that, with the "proletarianization of the senses" a fait accompli rather than a process to be reversed, he turns further toward the facticity of *hypomnemata* themselves (the order of material inscriptions). The arche-cinematic needs, he knows, to be other than metaphoric, a

modus of "consciousness" production as well as that of perceptual regimes: "'Consciousness' would then be this post-production center, this control room assembling the montage, the staging, the realization, and the direction, of the flow in primary, secondary, and tertiary retentions, of which the unconscious, full of protentional possibilities (including the speculative), would be the producer." [Stiegler 2010: 28] What need be reversed in the culture of this metaphorics, which includes by negative default the closet mimeticisms of cinema studies, is the narrative that a "cut" is introduced by the editing process of material "pieces of film" (Hitchcock), as if disruptively, rather than that "cut" pre-inhabiting linguistic consciousness (Cohen 2016).

20. The Western remainder of the anthropoid trajectory finds itself (justly) trolled from all sides in its geopolitical death by a thousand cuts and, inversely, superficial ubiquity. By Putin, clearly, but most notably in black-faced ISIS, which presents itself as a reverse anthropomorphism and death cult, out-miming internet web-sites, curtailing French TV stations, subsuming video-game imaginaries and HBO aesthetics (*Game of Thrones*). The beheading videos of Western faces mocks and plays the Western addiction to face and mimeticism. *ISIS* defaces—not only the ancient temples and cultural treasures but the tele-streaming totalization that posits itself as hyper-industrial modernity, "the West," yet whose teleological fantasies project escapist singularities. In this battle of the two nihilisms, the Western anthropos and his faceless, Ebola-like, decapitating, mass killing other—an "other" no Judith Butler could step in to empathize with—Žižek is disappointed in the *fake* "fundamentalism," seeing it as envious of what it rejects—even if that is the comfort and security and sexual ease of last man culture: "It may appear that the split between the permissive First World and the fundamentalist reaction to it runs more and more along the lines of the opposition between leading a long satisfying life full of material and cultural wealth and dedicating one's life to some transcendent cause. Is this antagonism not the one between what Nietzsche called "passive" and "active" nihilism? We in the West are the Nietzschean Last Men, immersed in stupid daily pleasures, while the Muslim radicals are ready to risk everything, engaged in the struggle up to their self-destruction" (Žižek, 2013). Gaining admiring analyses from cineastes, and having forced Obama to do the last thing he wanted to (open yet another Middle East war, particularly one the liquid and techno-evolutionary ISIS could, like Russia's "hybrid" war, simply win).

21. Let us leave aside whether "correlationism" is another name for a more subdued routine of the past (mimeticism, Heidegger's *Adequatio*, de Man's *descriptivity*, Benjamin's "historicism," et al.). Meillassoux stages this break with something of a dummy front, misleadingly familiar if hyperbolized as what it is not: the great error of 20th century confabulations was linguistic fetishism, like "deconstruction." Yet in positing what it definitively dismisses in opting for, well, totem words ("the absolute," "infinity," "randomness")

the power of the gesture as an assertion beyond the closing and closed order of inertial philosophemes and consumerist thought, he creates a hologram "unconscious" that leaks back—quite aside from miming the errata out of which something like deconstruction (arose around the materialities of semio-mnemonic "consciousness"). The closed orders of what lies outside the 21st philosophemic toolbox is not the absolute but climate change, bio-material mutations, the permeabilities of the inorganic, and the closing tele-imaginaries streamed incessantly that no magical "OOO" can claw down, even when moving productively to the peripheries (objects require subjects, even self-denying ones, and literary descriptives). Thus for Meillassoux, the return of the suppressed and denied is fairly immediate—as when the exemplar will be a close reading of Mallarme's *Un Coup de des*, or in the literalization of an *arche*-fossil that, purportedly, proves to the neo-philosophic saint (theology returns with a vengeance) that a world existed and exists and will exist outside of anthropos' "consciousness" (now largely managed and streamed), since that needed proof by way of a trace. Meillassoux, in short, rediscovers de Manian randomness and Benjaminian "prehistory," not to mention trace, while clothing his prose in swinging incense fumes. The desire to be free of fossils leads to a logic of future fossils (Barikin, 2004). It is not the "prison-house of language" that requires escaping (itself an ideologeme misreading Nietzsche for an imaginary left polemic) but the "prison-house" of generated referential, tele-political streams, mnemo-technic regimes, "correlationist" habits and conventions—the undead order of "tropological systems" digitally totalized and grammaticized.

22. "Face" is always *double tapped.* It must (it seems instantaneously) archive, recognize, then passively cite itself—rather like anything identified by "consciousness" as perception, experience, or even the "I." And like any identifiable perception ("phenomenalization"), is a composed and redacted citation—as a photograph is, despite its plead to index. Like the cinematic cut, which used to be marketed as a radical effect but now is recognized to barely catch up with "consciousness'" own mnemonic jump-cut editing of hourly rushes, the effect of coalescing or crediting a face of the friend, a core survival imperative of groups, seems instantaneous. Apparently any *Anthropocene era* will only be possible to name once this fellow, *anthropos,* is fully composite and displayed as an embedded algorithm. The problem is, perhaps, that the moment "anthropomorphism" becomes readable as other than a trope, as an invasive alien software that hides among and then freezes their ability to generate substitutive chains or transport—when it becomes, as it were, totalized—then any retirement of a culture of face implicitly involves the retirement of an entire story arc and trajectory of the very figural systems that manage cognition and knowledge bases. There would be not just a sedimentation of dead metaphor, but an evacuation of aura or mourning from figures that no longer function, are even figural, or perform their services. The good news might be that once you *double tap* face, you implicitly double

tap everything, which would appear to create an opening to reset without relapsing into the image of the *oikos* that, essentially, had performed as the cave, butcher shop, or reserve imaginary cavity in which interiorities would be proprietized, defended in advance, rendered a counter-chronotopic anchor or enclave, or what Wallace Stevens' calls "the jar." But if an era of face were retired what Derrida identified as *humanualism*—the determination of "human" difference by the technicity of the hand, going back to the cave painting—would also be closed with our being aware particularly. The not inglorious hand, and its ally the eye, are affiliates of face and adhere to the latter's mirage of a full spectrum *senses management center* (nose, tongue, ears, and so on), a head or *Capo*. Everything that then went by the name human, as an essential property, including *humanualism* in Derrida's formulation, but certainly including affect, identification, mimetic regimes, referentials, and mourning appear, rather than as core and defining traits (the heart, say), as the product of an artifacted and whirring citational circuit. It is less a question of a ghost in the machine than a machine that is the ghost. Of course, what is discretely implied is also the sudden retirement of the *humanualist* network, since the hand is now replaced by pixels and *digitalia*. Unlike the hand, which iconizes technics, touch, holding (conceptualizing), and proffers broad contact with body surfaces, fingers are multiplex, channel strikes and numbers, and are associated with mnemonic gadgets, and tap keys. But not only has the body not itself been told of this change, but during that gap of composition between cinematic cuts and the completed montage, narrative resin is secreted, and pasted back as a recuperated. The digital double taps "the body" and instantly knows how to counterfeit continuity—face identification, mimetic kazoos, and all this, the better to stream, harvest, and control. This seems obvious in the present, from corporate streams of climate denialists to Hollywood's "consciousness industry" (Stiegler). Hollywood, of course, has been quick to market climate disaster flicks in concert with corporate logics: Cli-Fi blockbusters and post-apocalyptic rants, mock-Biblical updates and earth-ending wormhole travel, and so on—each of which, oddly enough, allays anxiety in three ways: first, they all have survivors that regenerate a new future (it will be ok—well, someone will); second, they familiarize with all variety of catastrophe logics, so whatever arrives will remain familiar; and third, the public feel they've already seen the movie (even as they are subliminally instructed that "climate change" is to be ignored). With the politics of managed extinction well underway, with financial engineering at the forefront, "Hollywood" retains a privileged, entirely endemic, role.

23. This phantasmagoric essay returns, otherwise, to a core negotiation: a sentence fragment from "On Truth and Lies in an Extra-Moral Sense," a text fragment that (along with *Birth of Tragedy*) would define de Man's "Nietzsche" and, if you like, Nietzsche's "de Man"—the stripping away of all rhetorics of existential pathos before the "aesthetic" technogenesis of mnemonic power, material referential and perceptual regimes, "consciousness," and specifically

in rhetorical terms. What is overlooked is that this essay of Nietzsche's begins not as a scan of the tropological nature of words and dead words in cognitive tapestries but a questioning of the very role of "intellect" in the hominid trajectory—a pointing to where the automations of disembodied logics persist in relation to "life" (and against the backdrop of said creatures' momentary appearance on (and disappearance from) the planetary and cosmic stage. The preoccupation of the late de Man with *defacement* turns against a totalization of tropological systems—and the counter term that is left is shorn of any possible anthropomorphism itself, a "materiality of inscriptions": "Truth is a trope; a trope generates a norm or value; this value (or ideology) is no longer true. It is true that tropes are the producers of ideologies that are no longer true." There is something to be said today for a reading of world processes as complex "rhetorical" events caught within destructive vortices, reactive mimetologies, and pre-emptive maneuvers.

24. De Man tracked the privative import of zombie tropes but was reticent to assume any of the faux pathos of that impact of biomorphic life. He identified aesthetic ideology as what accelerates ecocide; he marked the cultural investment in "description" and "historicism" similarly; he deboned from the flesh of tropes a "materiality of inscription" that would be immaterial perceptually, and which would implacably extend into animation and life forms (*RNA*). He tagged *unreadability* yet did not bother to unfold what that implied literally—which is that no text is readable if it is not read through its gathering disclosure of ecocide. He refused all the "recuperative" options that welded the figure of light and the eye to the promise of knowing, and identified affect as a virtual affectation, but eschewed any prophetic pose. The ultimate reference to his trope of a "resistance to theory" was not the inscriptions out of which perceptual phantoms, historical narrative, referential grids, mnemo-technic regimes, mimetic ideologies, and ecocidal acceleration derive. The "theory" resisted is paralled oddly by the resistance to "climate change" today, which is the resistance to thinking this backloop at which an immateriality "materiality of inscription" occurs, and disappears into animation and biomorphic "life." These have now beached themselves in an "irreversibility" that penetrates reading.

25. Until recently, "climate change" was the perpetually arriving bad future to be warded off—a fiction that has settled in. That is how the passing of tipping points typically looks in an ecocidal civilization. After *tipping points* pass, the acceleration obeys its own vortex of negative feedback loops and sudden adjustments. But with tipping points past the temporal glue that held in place the discourse currency of *anthropos*—the temporal span of the "promise," about which credit, managed futures, and utopist lures spawn—snaps like the Fates' thread. A civilization that cannot promise even a "future" has abandoned—like the contract of face or tropological regimes, *humanualism* itself—the pretense of the "promise." Hence the shift today from a rhetoric of sustainability to one of "resilience."

26. Say, some HBO script titled "2015," in which a variety of oddities and events juggle for symptom status in a world in which biologemes and "reality" appear to have been hacked. In it, mega-droughts and destructive floodings accelerate, international "law" accords from the post-cold war parenthesis dissolve, democracies fall or falter systemically, and so on—genre stuff. But the twist either will sell or not: not only have tipping points quietly passed (and been discretely acknowledged by Western states), but as the fog of the 2008 economic "crisis" parted, or could afford to, what emerged to view was quite impressive: the greatest wealth transference, accomplished digitally, since Genghis, and without a shot fired by those harvested. But even this concealed another event in the plot—and here I am a bit suspicious, unsure if it will go directly to late night by its absurdity. Since this remarkable wealth transference was engineered by the same corporate class that streamed climate denialism to the masses, and since this same hyper-elite turned down acting at every opportunity, particularly the U.S., and iconically in Copenhagen, one begins to assume an outlandish Hollywood narrative is lurking, a narrative that is even disguised by public analysis, chitchat, and hand-wringing about "inequality"—another weasel term fallen from the Terracotta warrior army of tropes, like those of climate discourse, engineered for statuesque opacity. It certainly seems odd if the premise of Enlightenment rationalists, free market priests, and science itself assumed that with the proper evidence, or coming to light, of awareness of these horrific facts and these extinction friendly consequences, a coming together would occur to defend, well, self-interest, survival, and so on, which has turned out to be largely, if not emphatically, not the case. Rather, we found that after Copenhagen, they accelerated emissions. Now, of course, the same interests streaming climate change denialism (redundant cognitive paralysis, as practiced on Fox News to great effect), know and access more "deep state" and collapsing ecosystem reports than any "public"—so, first of all, they know. And the transference of hard assets to a new hyper-elite, and the "breakaway civilization" (and economy) enacted, is not an issue of "inequality" gone hyper again, an imbalance again in the capitalists' cycle to be wished to be returned to its norm (er, "equality"?), but the meme even in Piketty's hands (2014) serves its anaesthetizing purpose. But one of the pleasures of being past tipping points, which is to say irreversibly accelerant, is that one can set aside all cordial or considerate attempts to rally those addicted to its tinctures—ecologically-minded folk, honorable Cassandras, weak messianists (yes, still on auto-pilot), those seeking and proposing new "we's," or extended ones to accommodate some rupture of otherness or some yet unaccounted for thing or animeme outside of the pale, late, but now fully digitalized "anthropomorphic" apparatuses. And this makes me wonder if the script will fly: with this transference of terrestrial wealth and resources undertaken not just out of compulsive corruption and mega-acquisitiveness, as per Wall Street gaming itself, but with an eye to the predictions of agricultural collapse and population culling, such as made already in a 2003 climate change

report by the U.S. Department of Defense and C.I.A. (Townsend and Harris 2014). One had just witnessed, as no one notes, what will be understood in retrospect as a *species split*, and a delineation of disposable castes and territories going forward, and a hyper-elite at the cusp of exponential leaps in nanotechnologies, genetic engineering, robotics, bio-technics, hyper-militarized data security, and so on, willing to make the ultimate sacrifice of, in this imaginary, being the survivor caste of the species, those who would be genetically enhanced in several generations, and in control of remaining resources and so on, and look back on the messy residue of earlier "Anthropocene 3.0" types as a mop-up operation of those endangering species survival. You can round this out yourselves—but I count this screenplay as a robust sampling of climate comedy today. What an absurd, kitsch, and cynical script. But one does not always get to change channels. The issue with any totalization of cognitive systems is the tropological totalitarianism it lends itself to. So then, what does this "army of tropes" and its enforcers have to do with, well, passing tipping points today—and continuing to dissociate itself cognitively from the event? Any metaphorics of *light* and the angelic metaphor of the "human" and its interpretive recuperations appear, like Enlightenment templates, tools of a different order, the levers of hypo-scripts operating from where light and its others have not been metaphorized, blackness not invented nor "light" naturalized as solar god for the eye and life.

27. It would be an error to think of de Man as launching a mode of rhetorical reading ruled by the *combinatoire* of substitutive chains (anacoluthon, prosopopeia, metonomy, and so on, an enumeration Nietzsche's sentence he is rehearsing when interrupted). Rather, "tropes" would tend in late de Man to aggregate, appear finally all the same among themselves—the place of *doxa*, of "phenomenalization," of referential regimes. They are be displaced en masse by what, nonetheless, would have no access to perceptibility itself, an order of inscriptions.

28. It would seem, across this parenthesis related to the promise, one witnesses the cinematic migrations of the Greek *Autos* in advance of its successors and acquisitions—which includes the corporatization of the non-existent "we"—until, like the alien bursting from the host human's belly, it assembles itself autonomously, automatically, acquiring compensated elite humans as puppets and agents (lawmakers, lobbyists, the sociopathic hero CEO), on behalf of logics of exponential acceleration and extractivisms for which social or human realities are entirely irrelevant: this specter of an autonomous A.I. haunts the post-tipping point imaginary almost like a salvation—the inverse singularity concealed in the optimist versions. The trajectory anthropos would have cycled through from initiation to the corporatization of the person, in some regards "his" full exteriorization, might be cast as Grand Theft *Autos* (by itself).

29. For the rural folk of the U.S. state of Georgia who share a dwindling border river with their Tennessee neighbors, the latter are sub-humans (and vice versa)—as a segment of *The Daily Show* once documented hilariously. The video-game "Tribal War" simply turns on the annihilation of the enemy "tribe" (whichever), a variety of soccer hooliganism as *ISIS*. It is not accidental that the geo-political rift artifacted today between a "West" and an "East" that marks the de-legitimation of the former's universalist models, market hypocrisies, and "Enlightenment" legal facades levered by financial mafias rests not on the rift between post-communist societies and "the West" but a different history of writing systems and temporal and mnemonic settings, between nihilist monotheist alphabetacisms with teleological grammars and alternative hierarchies and definitions of human life (and economies of death).

30. For one of the many genealogies of the "future" offered today with the Janus-faced god of A.I. in view—the usual, salvation or annihilation, immortality or disposability—see Peter Holley 2015: "Apple co-founder on artificial intelligence: 'The future is scary and very bad for people'."

31. It would have been manifest in a constitutive resistance to reading *itself*, to reading "Plato" say, rather than transmuting and tagging that event as what produced memes the text explicitly disowns—for instance, the *eidos*, which in fact "Parmenides" flays on its conjecture from "young Socrates."

32. Perhaps a better prism for reading Nietzsche's "Rome" would be Michel Serres's interrogation of Rome's difference and self-erasing origin "legend"—producing the imperial techno-anthropoid void of philosophic originality, a remarkable reading of Rome's emergence from and as a "*black* box," a cave legend unlike Plato's and without sun (Serres 2015, 9-33).

33. Every time a Hitchcock (among others) marks the cinematic machine as the subject and predatory eye itself, pushing out or killing off mimetic human "stars" (from *The Lodger* on, yet most on display in *Psycho's* "mother" or the birds themselves), it is marked overtly.

34. When we survey, from the twenty-first century, how "the Greeks" would be edited, canonized, reduced to fingers pointing up and down, as by Raphael, it echoes what de Man finds occurs to Kant when passed through Schiller's ameliorative pop transcriptions—that nothing changed in centuries of interpretation once that would be installed. Or for "Plato," millennia. If the *eidos* never was posited in Plato's script as such, the artifice of the "Good" involved little more than an inverse transposition of an indigestible materiality of the mark itself, transposed across his oeuvre through the evolving figurations of "Socrates" (Plato's *Semiotik*, says Nietzsche), as hypogram, as *arete*, and so on. The helio-swathed "light" or sun was left for the hypocrite *lecteurs* Plato had so much trust for that he would exile, for them (no doubt chuckling), hearing or reading poets.

35. "On the one hand, a growing shadow of mass depoliticization is cast by such scientific studies of the Anthropocene as various networks of scientific and technical experts once again position themselves to administer from above and afar any collective efforts to mitigate or adapt to rapid anthropogenic climate change" (Luke 2015, 141).

36. https://www.jacobinmag.com/2015/03/anthropocene-capitalism-climate-change/

Chapter Three

37. See Miller 2010. I am grateful to *Derrida Today* (Edinburgh University Press) and to Nicole Anderson and Nick Mansfield for permission to reuse this essay in slightly changed form. See *Permissions* at the end of the book.

38. Jules Michelet, "Avenir! Avenir!"; cited from Walter Benjamin, *The Arcades Project*, 4. Percy Bysshe Shelley, "A Defence of Poetry," etc. I take these citations from Samolsky's admirable *Apocalyptic Futures* (Samolsky 2011).

39. See http://en.wikipedia.org/wiki/Philosopher's_stone, accessed May 31, 2009.

40. I perform this reading in *The Conflagration of Community: Fiction Before and After Auschwitz* (Chicago: University of Chicago Press, 2011) on the basis of hints in Kafka's own statements and in remarks by previous critics, as well as on the uncanny similarity between Kafka's narrative and later first-hand accounts of the Jews' transportation to Auschwitz, for example in Primo Levi's *Survival in Auschwitz*, Elie Wiesel's *Night*, or Imre Kertész's fictionalized account of his teenage experiences in the camps in *Fatelessness*.

41. See www.stoptrashingtheclimate.org. Accessed November 18, 2015.

42. The reader should note that the oral version of this paper included an accompanying PowerPoint presentation showing an array of dumps, garbage, and signs of melting worldwide, from the local dump on Deer Isle to the Antarctic ice-melt, a beach in Hawaii, and the top of Mt. Everest, where melting is exposing the debris of many climbing expeditions. The presentation also included some stills from the Pixar/Disney movie *WALL-E*, showing the whole world made uninhabitable by garbage everywhere. A large secondary literature on garbage exists. New evidence of global warming appears nearly every day in the media. An example is a recent short piece in *Science News* titled: "Oceans yield huge haul of plastic: 'Garbage patches' more common and deeper than thought." What is sometimes called "the Great Pacific Garbage Patch" is now, the article notes, measured as about twice the size of Alaska. "Some of these areas are like a black hole," Nikolai Maximenko, an oceanopgrapher at the University of Hawaii at Manoa, is quoted as having

said. "Once things are trapped there, they never escape." Most of this garbage is tiny bits of non-biodegradable plastic. (*Science News* [March 27, 2010], 8). Here are a few high points from the literature on garbage. I thank Dragan Kujundzic for identifying these for me. Jacques Derrida's "Biodegradables: Seven Diary Fragments" (*Critical Inquiry*, 15, 4 [Summer 1989], 812-73) is an aggressive answer to the six "responses" the editors of *Critical Inquiry* had recruited as comments on Derrida's "Paul de Man's War." It is also, however, an admirable exploration of the resonances echoing in those strange words "biodegradable" and "non-biodegradable." Derrida's essay is particularly forceful as a reflection on what is at stake when you ask whether a text can be the one or the other. Here are six books on garbage, waste, rubbish: Zygmunt Bauman, *Wasted Lives: Modernity and Its Outcasts* (Oxford: Polity, 2004); Gay Hawkins, *The Ethics of Waste: How We Relate to Rubbish* (Lanham, Maryland: Rowman & Littlefield, 2006); Susan Strasser, *Waste and Want: A Social History of Trash* (New York: Henry Holt, 1999); Heather Rogers, *Gone Tomorrow: The Hidden Life of Garbage* (New York: The New Press, 2005); William Rathje and Cullen Murphy, *Rubbish!: The Archaeology of Garbage* (New York: HarperCollins, 1992); Elizabeth Royte, *Garbage Land: On the Secret Trail of Trash* (New York: Little, Brown and Company, 2005). A series of brilliant articles and conference presentations by Natalka Freeland have explored "disposable culture" and "rubbish" in "the Victorian imagination," for example the dust heaps in Dickens's *Our Mutual Friend*.

43. In "Delegation: 'Deconstruction' *Contretemps*, Family Plots, and Climate Change," MS, 15.

44. Cited by from Foucault's *Power/Knowledge* by Giorgio Agamben in an essay on Foucault, "What Is an Apparatus," *What Is an Apparatus? And Other Essays*, trans David Kishik and Stefan Pedatella (Stanford, California: Stanford University Press, 2009), 2.

45. See Binaj Gurubacharya's essay at http://www.huffingtonpost.com/2009/05/25/appa-sherpa-warns-mount-e_n_207398.htm, accessed June 1, 2009.

Works Cited

Agamben, Giorgio. 1995. *Homo Sacer. Sovereign Power and Bare Life.* Trans. Daniel Heller-Roazen. Stanford: Stanfod University Press.

Agamben, Giorgio. 1999. *The Man Without Content.* Trans. Georgia Albert. Stanford: Stanford University Press.

Agamben, Giorgio. 2009. *What Is an Apparatus? And Other Essays.* Trans. David Kishik and Stefan Pedatella. Stanford, California: Stanford University Press.

Alaimo, Stacy. 2010. *Bodily Natures: Science, Environment, and the Material Self.* Bloomington: Indiana University Press.

Badiou, Alain. 2007. *Being and Event.* Trans. Oliver Feltham. London: Continuum.

Badiou, Alain. 2009. *Theory of the Subject.* Trans. Bruno Bosteels. London: Continuum.

Barikin, Amelia. 2014. "Arche-Fossils and Future Fossils: The Speculative Paleontology of Julian Charrière." In *Julian Charrière Future Fossil Spaces.* Ed. N. Schweizer. Musée cantonal des Beaux-arts in Lausanne & Mousse Publishing.

Barthes, Roland. 1972. *Mythologies.* Trans. Annette Lavers. London: MacMillan.

Bauman, Zygmunt. 2004. *Wasted Lives: Modernity and Its Outcasts.* Oxford: Polity.

Belsey, Catherine. 1980. *Critical Practice.* London: Methuen.

Benjamin, Walter. 1986. "Karl Kraus." in *Reflections: Essays, Aphorisms, Autobiographical Writing.* Ed. Peter Demetz. New York: Schocken Press, 432-458.

Bordo, Jonathan. 1992. "Ecological Peril, Modern Technology and the Postmodern Sublime." in *Shadow of Spirit: Postmodernism and Religion.* Ed. Philippa Berry and Andrew Wernick. London: Routledge, 1992: 165-178.

Butler, Judith. 1990. *Gender Trouble.* London: Routledge.

Chakrabarty, Dipesh. 2009. "The Climate of History: Four Theses," *Critical Inquiry* 35 (Winter 2009): 197-222.

Cheah, Pheng. 2009. "The Untimely Secret of Democracy." *Derrida and the Time of the Political*. Ed. Pheng Cheah and Suzanne Guerlac. Durham: Duke University Press, 2009. 74-96.

Clark, Nigel. 2011. *Inhuman Nature: Sociable Life on a Dynamic Planet*. London: SAGE, 2011.

Clark, Tim. 2015. *Ecocriticism on the Edge: The Anthropocene as a Threshold Concept.* London: Bloomsbury.

Cohen, Tom. 1998. *Ideology and Inscription: 'Cultural Studies' After Benjamin, de Man, and Bakhtin.* Cambridge: Cambridge University Press.

Cohen, Tom. 2009. "Tactless—the Severed Hand of JD." *Derrida Today* 2.1 (2009): 1-22.

Cohen, Tom. 2014. "The Telepolis—a Prehistory." *Affirmations*, 2.1 (2014): 85-109. http://affirmations.arts.unsw.edu.au/index.php?journal=aom&page=article&op=view&path%5B%5D=55&path%5B%5D=45.

Cohen, Tom. 2015. "Isis, Facelessness and Anthropomorphism: Thirteen Ways of Looking at ISIS—or, Jihadis 'R Us." Special Google Drive folio: The Semiotics of ISIS. Ed. Jonty Tiplady. November, 2015. https://drive.google.com/folderview?id=0BxyhPFYpy6g5Q09XdG1WVWFzRW8&usp=sharing

Cohen, Tom. 2016. "*Arche*-Cinema and the Politics of Extinction." *Boundary 2* (forthcoming).

Cohen, Tom, Claire Colebrook and J. Hillis Miller. 2012. *Theory and the Disappearing Future.* London: Routledge.

Copjec, Joan. 2002. *Imagine There's No Woman: Ethics and Sublimation*. Cambridge, Mass.: MIT Press.

Critchley, Simon. 1999. *The Ethics of Deconstruction*. Edinburgh: Edinburgh University Press.

Cubitt, Sean, Daniel Palmer and Nathaniel Tkacz. 2015. *Digital Light*. London: Open Humanities Press, 2015. http://openhumanitiespress.org/digital-light.html.

Dawkins, Richard. 2008. *The God Delusion*. Boston: Houghton Mifflin Co.

Day, Aidan. 2002. *Romanticism*. London: Routledge.

De Man, Paul. 1978. "The Epistemology of Metaphor." *Critical Inquiry*, 5.1 (Autumn 1978): 13-30.

De Man, Paul. 1979a. *Allegories of Reading*. New Haven: Yale University Press.

De Man, Paul. 1979b. "Autobiography as De-facement." *MLN*, 94. 5, Comparative Literature (December 1979): 919-930.

De Man, Paul. 1982. *Blindness and Insight: Essays in the Rhetoric of Contemporary Criticism*. Minneapolis: University of Minnesota Press.

De Man, Paul. 1984. *The Rhetoric of Romanticism*. New York: Columbia University Press.

De Man, Paul. 1996. *Aesthetic Ideology*. Ed. Andrzej Warminski. Minneapolis: University of Minnesota Press.

Deleuze, Gilles. 1994. *Difference and Repetition*. Trans. Paul Patton. New York: Columbia.

Deleuze, Gilles and Félix Guattari. 1987. *A Thousand Plateaus: Capitalism and Schizophrenia*. Trans. Brian Massumi. Minneapolis: University of Minnesota Press.

Dennett, Daniel C. 2006. *Breaking the Spell: Religion as a Natural Phenomenon*. New York: Viking.

Derrida Jacques. 1993. *Raising the Tone of Philosophy: Late Essays by Immanuel Kant, Transformative Critique*. Ed. Peter Fenves. Baltimore: Johns Hopkins University Press.

Derrida, Jacques, 1991. "At This Very Moment in This Work Here I Am." In Robert Bernasconi Ed. *Re-Reading Levinas*. London: Athlone. 11-50.

Derrida, Jacques. 1973. *Speech and Phenomena: And Other Essays on Husserl's Theory of Signs*. Evanston: Northwestern University Press.

Derrida, Jacques. 1978a. *Edmund Husserl's Origin of Geometry, An Introduction*. Trans. John P. Leavey, Jr. Lincoln: University of Nebraska Press.

Derrida, Jacques. 1978b. *Writing and Difference*. Trans. Alan Bass. London: Routledge.

Derrida, Jacques. 1982. *Positions*. Trans. Alan Bass. Chicago: University of Chicago Press.

Derrida, Jacques. 1984. "No Apocalypse, Not Now (Full Speed Ahead, Seven Missiles, Seven Missives):" Trans. Catherine Porter and Philip Lewis. *Diacritics*. 14.2 (Summer 1984): 20-31.

Derrida, Jacques. 1989. *Edmund Husserl's Origin of Geometry: An Introduction*. Trans. John P. Leavey Jr. Lincoln: University of Nebraska Press.

Derrida, Jacques. 2006. *Specters of Marx: The State of the Debt, the Work of Mourning and the New International*. Trans. Peggy Kamuf. London: Routledge.

Derrida, Jacques. 2008. *The Animal That Therefore I Am.* Trans. David Will. New York: Fordham.

Derrida, Jacques. 2009. *The Beast and the Sovereign* Trans. Geoffrey Bennington. Chicago: University of Chicago Press.

Derrida, Jacques. 2011. *The Beast and the Sovereign: Volume 2.* Trans. Geoffrey Bennington. Chicago: University of Chicago Press.

Derrida, Jacques. 2013. "The Fiction of the World." *The Oxford Literary Review.* 35.1 (2013): 1–3.

Diamond, Jared M. 1999. *Guns, Germs, and Steel: The Fates of Human Societies.* New York: Norton.

Eagleton, Terry. 1996. *Literary Theory: An Introduction.* Oxford: Blackwell.

Elden, Stuart. 2008. "Eugen Fink and the Question of the World." *Parrhesia* 5 (2008) 48-59.

Fink, Eugen. 1970. "The Phenomenological Philosophy of Edmund Husserl and Contemporary Criticism." in R.O. Elveton Ed. *The Phenomenology of Husserl: Selected Critical Readings.* Chicago: Quadrangle, 1970: 74-147.

Gasche, Rodolphe. 2009. *Europe, Or the Infinite Task: A Study of a Philosophical Concept.* Stanford University Press.

Gould, Stephen Jay. 1998. "Scale Models." http://www.forbes.com/asap/1998/1130/157.html. Accessed October 20, 2015.

Gratton, Peter. 2013. "Post-Deconstructive Realism: It's About Time." *Speculations: A Journal of Speculative Realism* IV (2013): 84-90.

Habermas, Jürgen. 1990. *Moral Consciousness and Communicative Action.* Translated by Christian Lenhardt, Shierry Weber Nicholsen. Cambridge, Mass.: MIT Press.

Hägglund, Martin. 2008. *Radical Atheism: Derrida and the Time of Life.* Stanford: Stanford University Press.

Hamilton, Clive. 2012. "Utopias in the Anthropocene." Plenary session of the American Sociological Association, Denver, 17 August 2012. http://mahb.stanford.edu/wp-content/uploads/2012/08/2012-Clive-Hamilton-Denver-ASA-Talk.pdf. Accessed 6/10/2015.

Hamilton, Clive. 2013a. *Earthmasters: The Dawn of the Age of Climate Engineering.* New Haven: Yale University Press.

Hamilton, Clive 2013b. "Climate change signals the end of the social sciences http://theconversation.com/climate-change-signals-the-end-of-the-social-sciences-11722

Hamilton, Clive. 2015. "Getting the Anthropocene So Wrong." *The Anthropocene Review*. (2015): 1-6.

Haraway, Donna. 2015. "Anthropocene, Capitalocene, Plantationocene, Chthulucene: Making Kin." *Environmental Humanities*, 6 (2015):159-165.

Hardt, Michael and Antonio Negri. 2000. *Empire*. Cambridge, Mass.: Harvard University Press.

Hawkins, Gay. 2006. *The Ethics of Waste: How We Relate to Rubbish*. Lanham, Maryland: Rowman & Littlefield.

Hedges, Chris. 2013. "The Pathology of the Rich—Chris Hedges on Reality Asserts Itself (interview)." *The Real News*, December 5, 2013. http://therealnews.com/t2/index.php?option=com_content&task=view&id=31&Itemid=74&jumival=11150; http://progressive.org/chris_hedges_interview.htm. Accessed 5/25/2015.

Hitchens, Christopher. 2007. *God is not Great: How Religion Poisons Everything*. New York: Twelve.

Holley, Peter. 2015. "Apple co-founder on artificial intelligence: 'The future is scary and very bad for people.'" *Washington Post*, March 23, 2015. http://www.washingtonpost.com/blogs/the-switch/wp/2015/03/24/apple-co-founder-on-artificial-intelligence-the-future-is-scary-and-very-bad-for-people/. Accessed 4/23/2015.

Irigaray, Luce 2002. *Between East and West*: From *Singularity to Community*. Trans. Stephen Pluháček. New York: Columbia University Press.

Irigaray, Luce. 1985. *Speculum of the Other Woman*. Trans. Gillian C. Gill. Ithaca: Cornell University Press.

Jameson, Fredric. 1977. "Imaginary and Symbolic in Lacan: Marxism, Psychoanalytic Criticism, and the Problem of the Subject." *Yale French Studies* 55/56, (1977): 338-395.

Jameson, Fredric. 1981. *The Political Unconscious: Narrative as a Socially Symbolic Act*. Ithaca: Cornell University Press.

Jameson, Fredric. 2005. *Archaeologies of the Future: The Desire Called Utopia and Other Science Fictions*. Durham: Duke University Press.

Joyce, James. 1964. *A Portrait of the Artist as a Young Man*. New York: Viking.

Kant, Immanuel. 1997. *Critique of Practical Reason*. Ed. Mary Gregor. Cambridge: Cambridge University Press.

Keats, John. n.d. *Bright star, would I were stedfast as thou art*. *Poetry Foundation*. Poetry Foundation. Accessed November 20, 2015. http://www.poetryfoundation.org/poem/173733#poem

Klein, Naomi. 2014. *This Changes Everything: Capitalism vs. the Climate*. New York: Simon and Schuster.

Kolbert, Elizabeth. 2009. "The Sixth Extinction?: There have been five great die-offs in history. This time, the cataclysm is us." *The New Yorker*, May 25, 2009, 52-63.

Lacan, Jacques. 1991. *The Seminar of Jacques Lacan: Freud's Papers on Technique*. Ed. Jacques-Alain Miller, Trans. John Forrester. New York: Norton.

Latour, Bruno. 2011. "Waiting for Gaia. Composing the Common World Through Arts and Politics." http://www.bruno-latour.fr/sites/default/files/124-GAIA-LONDON-SPEAP_0.pdf

Latour, Bruno. 2014a. "On Some of the Affects of Capitalism." http://www.bruno-latour.fr/sites/default/files/136-AFFECTS-OF-K-COPENHAGUE.pdf Accessed October 22, 2015.

Latour, Bruno. 2014b. "War and Peace in an Age of Ecological Conflicts." http://www.bruno-latour.fr/sites/default/files/130-VANCOUVER-RJE-14pdf.pdf

Lukács, György. 1963. *The Historical Novel*. Trans. Hannah and Stanley Mitchell. Boston: Beacon Press.

Lyotard, Jean-Francois. 1997. *Postmodern Fables*. Trans. Georges Van Den Abbeele. Minneapolis: University of Minnesota Press.

Manahan, Stanley. 2014. *Sustainocene: Managing the Anthrosphere for Sustainability in the Anthropocene Epoch*. Columbia: ChemChar Research.

Marche, Stephen. 2015. "The Epidemic of Facelessness." *New York Times*, February 14, 2015. http://www.nytimes.com/2015/02/15/opinion/sunday/the-epidemic-of-facelessness.html. Accessed 3/21/2015.

Massumi, Brian. 2014. *What Animals Teach Us About Politics*. Durham: Duke University Press.

Maturana, Humberto R. and Francisco J. Varela. 1987. *The Tree of Knowledge: The Biological Roots of Human Understanding*. Boston: New Science Library.

Meillassoux, Quentin. 2008. "Spectral Dilemma" *Collapse IV* (2008): 261-276.

Meillassoux, Quentin. 2010a. *After Finitude: An Essay on the Necessity of Contingency*. Continuum.

Meillassoux, Quentin. 2010b. "The Immanence of the World Beyond," in Connor Cunningham & Peter M. Candler Ed. *The Grandeur of Reason*. SCM Press, 2010: 444-478.

Meillassoux, Quentin. 2012. *The Number and the Siren: A Decipherment of Mallarmé's* Coup de Dés. Trans. Robin MacKay. Falmouth: Urbanomic, 2012.

Miller, J. Hillis. 2010. "Anachronistic Reading," in *Derrida Today*, 3 (May 2010): 75-91.

Miller, J. Hillis. 2015. "Mixed Media Forever: The Internet as Spectacle; or, The Digital Transformation of Literary Studies." *An Innocent Abroad*. Evanston, Illinois: Northwestern University Press, 2015: 243-256.

Moore, Jason. 2014. "The Capitalocene, Part II: Abstract Social Nature and the Limits to Capital" and "The Capitalocene, Part I: On the Nature & Origins of Our Ecological Crisis." http://www.jasonwmoore.com/Essays.html

Morton, Timothy. 2007. *Ecology without Nature: Rethinking Environmental Aesthetics*. Cambridge, Mass.: Harvard University Press.

Morton, Timothy. 2013. *Hyperobjects: Philosophy and Ecology After the End of the World*. Minneapolis: University of Minnesota Press.

Morton, Timothy. 2015. "This Biosphere Which Is Not One: Towards Weird Essentialism." *Journal of the British Society for Phenomenology*. http://dx.doi.org/10.1080/00071773.2014.960745.

Myers, Norman and Andrew H. Knoll. 2001. "The biotic crisis and the future of evolution," in *Proceedings of the National Academy of Sciences* (98.10, May 8, 2001): 5389-5392. http://www.pnas.org/content/98/10/5389.full. Accessed 6/10/2015.

Naas, Michael. 2013. "Derrida at the Ends of the World." In Amy Swiffen and Joshua Nichols Eds. *The Ends of History: Questioning the Stakes of Historical Reason*. New York: Routledge. 161-178.

Nagel, Thomas. 1974. "What is it Like to Be a Bat." *The Philosophical Review* LXXXIII, 4 (October 1974): 435-50.

Nietzsche, Friedrich. 1995. *Unfashionable Observations*. Trans. Richard T. Gray. Stanford: Stanford University Press.

Nietzsche, Friedrich. 2005. *The Anti-Christ, Ecce Homo, Twilight of the Idols: And Other Writings*. Ed Aaron Ridley and Judith Norman. Trans. Judith Norman. Cambridge: Cambridge University Press.

Nietzsche, Friedrich. 2009. *Writings From the Early Notebooks*. Ed. Raymond Geuss and Alexander Nehamas. Trans. Ladislaus Lob. Cambridge: Cambridge University Press.

Parikka, Jussi. 2015. *The Anthrobscene*. Minneapolis: University of Minnesota Press.

Piketty, Thomas. 2014. *Capital in the 21s Century*. Trans. Arthur Goldhammer. Belknap Press.

Pirici Alexandra and Raluca Voinea. 2015. "Manifesto for the Gynecene." http://infinitexpansion.net/gynecene/

Protevi, John. 2009. *Political affect: connecting the social and the somatic*. Minneapolis: University of Minnesota Press.

Rathje, William and Cullen Murphy. 1992. *Rubbish!: The Archaeology of Garbage*. New York: HarperCollins.

Rogers, Heather. 2005. *Gone Tomorrow: The Hidden Life of Garbage*. New York: The New Press.

Royte, Elizabeth. 2005. *Garbage Land: On the Secret Trail of Trash*. New York: Little, Brown and Company.

Rugh, Peter. 2013. "Learning to Live in the Anthropocene." http://www.occupy.com/article/learning-live-anthropocene

Samolsky, Robert. 2011. *Apocalyptic Futures: Marked Bodies and the Violence of the Text in Kafka, Conrad and Coetzee*. New York: Fordham University Press.

Scranton, Roy. 2013. "Learning How to Die in the Anthropocene." http://opinionator.blogs.nytimes.com/2013/11/10/learning-how-to-die-in-the-anthropocene/?_r=0

Serres, Michel. 2015. *Rome: The First Book of Foundations*. Trans. Randolph Burks. London: Bloomsbury.

Sloterdijk, Peter. 2013. *You Must Change Your Life*. Cambridge: Polity.

Stevens, Wallace. 2006. *Collected Poems*. London: Faber and Faber.

Stiegler, Bernard. 2008. *Technics and Time: Disorientation 2*. Trans. Stephen Barker. Stanford: Stanford University Press.

Stiegler, Bernard. 2010. *Technics and Time, 3: Cinematic Time and the Question of Malaise*. Trans. Stephen Barker. Stanford University Press.

Stiegler, Bernard. 2011. *Decadence of Industrial Democracies*. Trans. Daniel Ross. Cambridge: Polity.

Stiegler, Bernard. 2013a. "The Organology of Dreams and *Arche*-Cinema," *Screening the Past*, June 2013. http://www.screeningthepast.com/2013/06/the-organology-of-dreams-and-arche-cinema/

Stiegler, Bernard. 2014. *The Lost Spirit of Capitalism: Disbelief and Discredit*, 3. Trans. Daniel Ross. Polity Press.

Stiegler, Bernard. 2015. "Escaping the Anthropocene." https://www.academia.edu/12692287/Bernard_Stiegler_Escaping_the_Anthropocene_2015_ Accessed. October 22, 2015.

Strasser, Susan. 1999. *Waste and Want: A Social History of Trash*. New York: Henry Holt.

Truffaut. Francois and Helen G. Scott. 1963. *Hitchcock*. New York: Simon and Schuster, 1963.

Wark, McKenzie. 2015. *Molecular Red: Theory for the Anthropocene*. London: Verso.

Wiegman, Robyn. 2014. "The Times We're In: Queer Feminist Criticism and the Reparative 'Turn.' *Feminist Theory* 15.1 (April 2014): 4-25.

Žižek, Slavoj. 2014. "ISIS Is a Disgrace to True Fundamentalism." *New York Times, September* 3, 2014. http://opinionator.blogs.nytimes.com/2014/09/03/isis-is-a-disgrace-to-true-fundamentalism/. Accessed 5/12/2015.

Permissions

Chapter Three

J. Hillis Miller is grateful to *Derrida Today* (Edinburgh University Press) and to Nicole Anderson and Nick Mansfield for permission to reuse "Anachronistic Reading," in *Derrida Today,* vol. 3 (May 2010), 75-91, in slightly changed form.

J. Hillis Miller is also grateful to Northwestern University Press and to Liz Hamilton for permission to reuse in revised form the essay "Mixed Media Forever: The Internet as Spectacle; or, The Digital Transformation of Literary Studies." Here is the required copyright statement: Adapted from *Am Innocent Abroad.*

Lightning Source UK Ltd.
Milton Keynes UK
UKHW04f0619190918
329155UK00001B/420/P

9 781785 420153